174.2 Bol
Ending and extending life
Boleyn-Fitzgerald, Miriam.

WITHDRAWN

CONTEMPORARY ISSUES IN SCIENCE

ENDING AND EXTENDING LIFE

CONTEMPORARY ISSUES in SCIENCE

ENDING AND EXTENDING LIFE

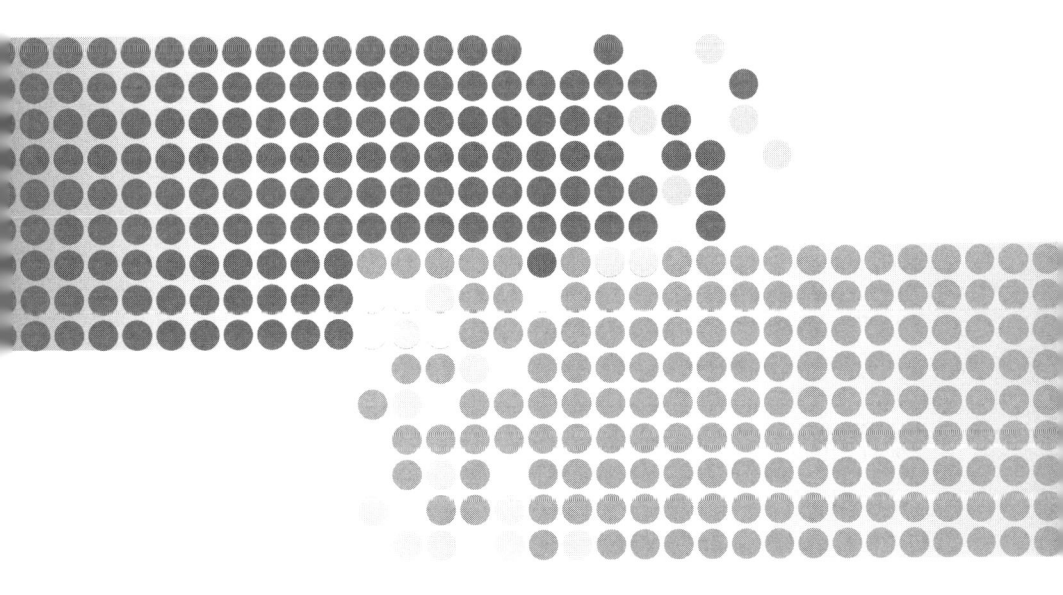

MIRIAM BOLEYN-FITZGERALD

WABASH CARNEGIE PUBLIC LIBRARY

✓ Facts On File
An imprint of Infobase Publishing

ENDING AND EXTENDING LIFE

Copyright © 2010 by Miriam Boleyn-Fitzgerald

All rights reserved. No part of this book may be reproduced or utilized in any form or by any means, electronic or mechanical, including photocopying, recording, or by any information storage or retrieval systems, without permission in writing from the publisher. For information contact:

Facts On File, Inc.
An imprint of Infobase Publishing
132 West 31st Street
New York NY 10001

Library of Congress Cataloging-in-Publication Data
Boleyn-Fitzgerald, Miriam.
  Ending and extending life / Miriam Boleyn-Fitzgerald.
    p. cm. — (Contemporary Issues in Science)
  Includes bibliographical references and index.
  ISBN-13: 978-0-8160-6205-8
  ISBN-10: 0-8160-6205-6
  1. Medical ethics. 2. Medical innovations—Moral and ethical aspects.
I. Title.
  R724.F26 2009
  174.2—dc22   2008030547

Facts On File books are available at special discounts when purchased in bulk quantities for businesses, associations, institutions, or sales promotions. Please call our Special Sales Department in New York at (212) 967-8800 or (800) 322-8755.

You can find Facts On File on the World Wide Web at http://www.factsonfile.com

Excerpts included herewith have been reprinted by permission of the copyright holders; the author has made every effort to contact copyright holders. The publishers will be glad to rectify, in future editions, any errors or omissions brought to their notice.

Text design and composition by Annie O'Donnell
Illustrations by Melissa Ericksen and Jeremy Eagle
Photo research by Tobi Zausner, Ph.D.
Cover printed by Art Print, Taylor, Pa.
Book printed and bound by Maple-Vail Book Manufacturing Group, York, Pa.
Date printed: February, 2010
Printed in the United States of America

10 9 8 7 6 5 4 3 2 1

This book is printed on acid-free paper.

CONTENTS

Preface	viii
Acknowledgments	xiii
Introduction	xiv

1 ETHICAL PRINCIPLES IN MEDICAL RESEARCH — 1

The Tuskegee Syphilis Study	1
Ethical Principles: Nuremberg to Belmont	8
Institutional Review Boards	11
The HIV Vertical Transmission Trials	15
Summary	22

2 VULNERABLE POPULATIONS — 23

Research with Children: The Willowbrook Hepatitis Study	25
Protections for Children	27
The Kennedy Krieger Institute Lead-Paint Study	29
Lead-Based Paint in Homes	33
Research with Emergency Room Patients: Ebb Cade and the Plutonium Injections	34
Protections for Emergency Room Patients	37
PolyHeme: The Synthetic Blood Trial	38
Summary	45

3 RESEARCH WITH ANIMALS — 46

Ethical Positions on Animal Research	50
Harry Harlow and Attachment Theory	54

OncoMouse   55
Protections for Animal Subjects   61
Xenotransplantation   63
Summary   72

4 WHEN LIFE ENDS   74
Karen Quinlan   74
Defining Death   76
Extend Life or Hasten Death?   85
Theresa Schiavo   87
Living Wills and Medical Proxies   91
Summary   95

5 LIFE-EXTENDING TECHNOLOGY   98
The Mechanical Ventilator   98
Tamoxifen: The First Chemopreventive Drug for Cancer   102
Tube Feeding   106
Neuroimaging   109
Organ Transplants   114
Summary   120

6 LIFE EXTENSION, AGING, AND PALLIATIVE CARE   122
The Science of Life Extension   123
Ethical Positions on Extending the Human Life Span   125
Life Expectancy at the Start of the 21st Century   126
Long-Term Care   128
The Growing Global Burden of Chronic Disease   129
Hospice and Palliative Care   132
Summary   134

7 NEW TECHNOLOGY AND THE COST OF TREATMENT   136
The Rising Costs of Health Care   136
Health Care Coverage in the United States   139

Costs of New Technology	140
The HIV/AIDS Pandemic: Disparities in Distribution of Care	146
U.S. Funding for HIV Treatment and Prevention	148
Summary	156
8 HEALTH, DISEASE, AND WELLNESS	**158**
Defining Health and Disease	158
Contemporary Debates: What Is Fair, and What Counts as Treatment?	161
Drugs to Treat Childhood Behavior	170
Summary	172
Chronology	174
Glossary	180
Further Resources	194
Index	213

PREFACE

"Whenever the people are well-informed, they can be trusted with their own government. Whenever things get so far wrong as to attract their notice, they may be relied on to set them to rights."

—Thomas Jefferson

In today's high-speed, high-pressure world, keeping up with the latest scientific and technological discoveries can seem an overwhelming, even impossible task. Each new day brings a fresh batch of information about how the world works; how human bodies and minds work; how human civilization can "work" the world by applying its collective knowledge. Switch on a television news program or the Internet at this very moment—pick up any newspaper or current interest magazine—and stories about health and the environment, worries about national security and violent crime, or advertisements for the latest communication and entertainment gadgets will abound.

Given the nonstop flow of information and commercial pressures, it may seem that a surface understanding of scientific and technological issues is the only realistic goal. The Contemporary Issues in Science set is designed to dispel the myth that a deeper understanding of new findings in science and technology—and therefore considerable power to influence their use—is out of reach of nonspecialists and should be "left to the experts." The set reviews current topics of universal relevance like global warming, conservation, weapons of mass destruction, genetic engineering, medical research ethics, and life extension and explores—through the lens of real people's stories—how recent discoveries have changed daily life and are likely to alter it in the future.

Preface

Stories featured in the set have received attention in the popular press—often provoking heated controversy at a local, national, and sometimes international level—because beneath the headlines lie sticky questions about how new knowledge should, or should not, be applied, as illustrated by the following examples:

- *Genetic engineering.* The pace of discovery about the human genome and the genomes of other animal and plant species has been breathless since the year 1953, when James Watson and Francis Crick first described the double helix structure of deoxyribonucleic acid (DNA), the chemical substance that acts as a blueprint for building, running, and maintaining all living organisms. In April 2003—a mere 50 years later—sequencing of the human genome was complete. This impressive surge in knowledge about our genes has been accompanied by intense hopes—and intense fears—about newfound technical powers to manipulate the production of life. The tragic death of 18-year-old Jesse Gelsinger in a 1999 gene therapy trial begged obvious questions: Can medical investigators ever obtain truly informed consent from a volunteer when the risks of an experimental procedure are largely unknown? Are the potential benefits of gene therapy worth the unknown public-health risks of altering the human genome using viral vectors? What are the environmental risks of creating transgenic plant and animal species?
- *End-of-life care.* Bold medical innovations like mechanical ventilation, organ transplantation, and tube feeding have saved and improved the lives of millions of patients since the 1950s. A state of profound unconsciousness known as "irreversible coma" first occurred with the ventilator; before its availability, patients without working respiratory systems died from lack of oxygen. Now the bodies of severely brain-damaged and brain-dead people can be maintained indefinitely

with a steady supply of oxygen to their living tissues. Theresa Schiavo's case—and other controversial end-of-life cases—shows how loved ones and medical professionals try to grapple with agonizing questions like: When are medical interventions extending meaningful life, and when are they inappropriately prolonging death? If a patient's wishes cannot be known with certainty, who should decide her fate?

- *Consumer choice.* Using cheap and plentiful energy; selecting personal transportation modes; building and occupying homes; consuming . . . well, just about anything: These options are all realized through technological innovation. Consumer choice is credited for dramatically improving quality of life in North America over the past century, but it has also created a suite of forbidding problems: global climate change, pollution, urban sprawl, and resource depletion. Can modern consumers—especially the rapidly increasing Chinese and Indian "middle-classes"—enjoy the same choices, or the same quality of life, as North Americans of the last half of the 20th century? Will purely technological solutions for problems arise (e.g., will a form of cheap and reliable carbon sequestration be developed to store carbon dioxide, allowing coal to be used to produce cleaner electricity)? Or will technology provide the means for a dramatic change in how people live and work (e.g., will ubiquitous broadband and wireless access lead to the delocalized office—employees always at work, so there is no need to "go to work," no matter where they are)?

- *Water.* With "peak water" (the maximum amount of clean, usable water available globally) predicted to occur sometime in the next 25 years, this vital natural resource is certain to be the source of national and regional conflicts. Water plays an essential role not only in living processes but in industrial-scale heating and cooling and in new alternative energy technologies such as coal gasification, hydrogen production,

Preface

and biofuels conversion. Water also figures highly in global climate change, acting both as a greenhouse gas and as a dynamic heat reservoir. For humankind's clean water requirements, is technological advancement the problem or is it the solution? Will gigantic energy-efficient desalination plants turn countries with ocean coastlines into the new "wet" OPEC, with "clear gold" (water) replacing "black gold" (petroleum) as the preeminent wealth-generating natural resource? Can technological innovation lessen the terrible toll that floods and droughts take on property and human lives?

- *Privacy.* Today, all bits and pieces of personal information—financial, medical, political, religious, identity-by-association, consumer preference, and lifestyle—are being collected, parsed, amalgamated, mined, and analyzed at a rate, and to an extent, unimaginable a decade ago. An individual's personal information can be collected, shared, exchanged, sold, disseminated, and broadcast without notice given to, or permission received from, the individual—and all perfectly legally. Identity theft is a widespread and growing problem—a phenomenon both created and addressed by modern electronic and software technologies. The use of e-mail to acquire personal financial information under false pretenses, known as "phishing," was estimated to have cost U.S. citizens over $2.8 billion in 2006. Can the benefits of instantaneous and remote transactions—financial, consumer-based, social, and educational—ever outweigh the loss of privacy or the risk of being victimized? Who really owns a person's digital identity—the individual, banks, insurance companies, or government agencies?

- *Weapons.* On August 6, 1945, the city of Hiroshima, Japan, was annihilated by an atomic bomb that killed an estimated 70,000 civilians instantly. Radio Tokyo described the extent of the devastation in a broadcast intercepted by Allied forces: "Practically all living things, human and animal, were literally seared to death." Three days later, a second nuclear bomb was detonated—this time over the southern port city of Nagasaki—killing another 40,000 to 75,000 people. Nuclear weapons have not been used since, but many countries have sought and achieved the technology to deploy them. What is the real threat of nuclear warfare in the early 21st century? What other potentially devastating weapons are being developed today, and how can human civilization avoid its own violent destruction?

Whether readers are students considering a career in a scientific or technical field, science or social studies teachers or librarians, or inquisitive people of any age with personal, professional, or political interests in how new knowledge is applied, the Contemporary Issues in Science set places fresh research findings in the context of real-life stories, clarifying the technical and ethical subtleties behind the headlines and supporting an engaged, informed citizenry.

ACKNOWLEDGMENTS

I would like to extend special thanks to the creative team behind these books: Frank Darmstadt, executive editor, for his keen editorial eye; Jodie Rhodes, literary agent, for making the match; Tobi Zausner, photo researcher, for her talent and tenacity in hunting down pictures; Melissa Ericksen, for her outstanding work on illustrations; Annie O'Donnell, for designing a visually engaging text: Alexandra Lo Re, for her meticulous readings; and Peter Faguy, for his shared vision and passion.

It is with immense gratitude that I dedicate the set to my husband, Patrick, and my son, Aidan, who fuel my daily writing process with a potent blend of patience, unconditional love, and good humor. Aidan—who produces approximately 100 books in the span of time it takes his mom to produce one (but tries hard not to make me feel bad about it)—is my reason for everything. Patrick—whose careful readings and quick insights have improved the clarity of drafts (not to mention the clarity of his wife)—makes everything possible.

I owe an enormous debt of thanks to the teachers and professional mentors who helped me learn about science, ethics, and policy, and—most influentially—how to talk and write about them so that people might care. The list is too long for this page, but it includes John Boccio, Diarmuid Maguire, Amy Bug, Kevin Knobloch, Ruth Faden, Sarah Flynn, Dan Guttman, Anna Mastroianni, Jeff Kahn, Gil Whittemore, Charlie Weiner, Tom Cochran, Stan Norris, Jacob Scherr, Harriet Ritvo, Hugh Gusterson, Leo Marx, Jill Conway, Deborah Fitzgerald, Evelyn Fox Keller, and finally my parents—Alyce and Ken—who first taught me that science is for girls.

INTRODUCTION

In 1952, a 12-year-old girl was dying in a hospital in Copenhagen, Denmark. A victim of the polio epidemic raging throughout Northern Europe at the time, the girl was paralyzed, her left lung had collapsed, and she had entered respiratory failure. Doctors placed her inside a tank ventilator, or "iron lung," in an attempt to help her breathe. The iron lung functioned by pumping air out of the tank around her to allow her chest to expand and her lungs to partially fill, but it was not enough. She was not getting the oxygen she needed.

The anesthesiologist Bjørn Ibsen saw her, and he did something extraordinary: He inserted a tube down her windpipe and attached a bag he normally would have used to administer anesthesia during surgery. He began to squeeze the bag rhythmically to force air into her lungs, and she survived.

Ibsen's lifesaving improvisation spread like wildfire. Patients in respiratory failure from polio were treated in Copenhagen by teams of nurses and medical students who took turns squeezing bags to keep victims alive. That year also saw the worst outbreak of polio on record in the United States, and soon Massachusetts General Hospital was using Ibsen's ingenious method to sustain their polio patients. A machine was quickly devised to replace the human power of nurses and students, and the modern mechanical ventilator was born.

Prior to the 1950s, respiratory failure was, in all but extraordinary cases, a death sentence. Now, thanks to Ibsen's quick, creative thinking, there was a new way to sustain lives while patients recovered from illness or injury. There was also a new need for special wards and "intensive care" techniques to treat patients who were being kept alive by mechanical ventilation, and families and doctors suddenly faced new, often agonizing

Introduction

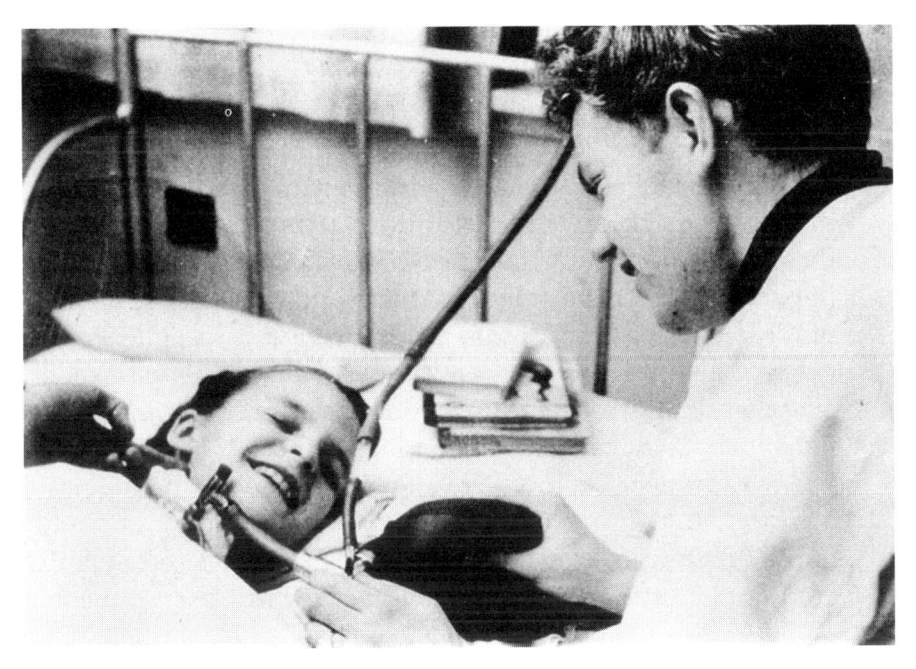

Bjørn Ibsen's life-saving manual ventilation system: A medical student squeezes a bag, delivering oxygen to a young girl's lungs through a cuffed tracheotomy tube. *(© 2007 American Thoracic Society)*

decisions about when to withdraw care from a loved one who would not recover.

Ending and Extending Life takes a detailed look at several bold medical innovations that have saved and improved the lives of millions of patients since Ibsen's time—mechanical ventilation, organ transplantation and other lifesaving surgical techniques, new diagnostic technologies (such as CT and MRI scans), advanced chemotherapy and antiviral drugs, and tube feeding—and their exciting and complex implications for patients, loved ones, and health care practitioners.

Medical innovation requires research, and research depends on patients and healthy volunteers willing to assume risks. *Ending and Extending Life* also considers ethical decisions faced by medical researchers on the path to discovery, using real patients' stories to highlight key ethical principles and their application to the everyday practice of medicine.

Each story chosen for the book has received attention in the popular press and has provoked controversy at a local, national, or even international level. Yet at the heart of each of these stories, underneath the political rhetoric and media hype, are the lives of real patients. Students considering careers in health care, medical research, and medical ethics can immerse themselves in these stories to better grasp the ethical choices involved and the importance of weighing and balancing potential consequences of those choices carefully, on a case-by-case basis. The Tuskegee syphilis study is a famous example of medical research gone awry; while Bjørn Ibsen's invention was working miracles for polio victims all over the United States and Northern Europe, U.S. government doctors were allowing 400 African-American men to live with untreated syphilis for the purpose of studying the natural course of this disease. (The Tuskegee syphilis study and its aftermath is the subject of the first part of chapter 1.) Revelations about this ethically egregious experiment, and about others involving institutionalized children, prisoners, and other vulnerable populations, propelled ethical questions to the forefront of the national consciousness in the early 1970s, and medical ethics became an academic discipline and profession in its own right.

At the research end of the process, medical ethics is concerned with balancing the risks and potential benefits of experimentation, and ensuring that human subjects are truly informed and protected throughout the process. When medical research is successful, it creates knowledge, paves the way for future innovation, and leads to longer, healthier lives. At the treatment stage, medical ethics grapples with issues like when, if ever, medical care should be withdrawn from a dying patient or from a patient lacking conscious awareness; who should decide a patient's fate when her wishes cannot be known; and whether equal access to care is a right or a privilege, a national or a global issue.

These questions infuse the real people's stories in this volume. They are as sticky as they are because they reveal people's scientific, ethical, and spiritual intuitions about what it means to be alive—and what it means to live a life of value—and because those intuitions can often come into conflict.

Ending and Extending Life follows the development and interplay of research ethics and of major advances in life-extending treatment from the mid-20th through the early 21st centuries. Chapter 1 traces the development of ethical codes for medical research since the Nazi experiments of World War II, and chapter 2 considers special issues arising with research involving vulnerable populations. Chapter 3 looks at medical research using animals—including the rapidly growing field of cross-species transplantation (xenotransplantation)—and at a range of perspectives on the moral status and treatment of animals.

Chapter 4 examines the high-profile end-of-life cases of Karen Quinlan and Theresa Schiavo and at how changing medical realities have been integrated into clinical and theoretical definitions of death. Chapter 5 takes a closer look at four life-extending technologies—mechanical ventilation, tube feeding, neuroimaging, and organ transplantation—and at special issues that arise with each in the context of four life-and-death stories.

The final three chapters explore the broader social and cultural contexts within which medical discoveries have emerged, and which, in turn, have been shaped by these discoveries. Chapter 6 looks at increasing life expectancy nationally and globally, how some demographic groups have benefited from these trends more than others, and how long-term medical care has adjusted to the needs of an aging population. Chapter 7 looks at some of the causes and consequences of rapidly rising medical costs, while chapter 8 looks at ways in which mainstream cultural assumptions about health and disease can determine which conditions—and which patients—receive medical treatment.

Tensions between ideas about what is right and the practical challenges inherent in the real-world practice of medicine will continue to generate fruitful debate as new medical discoveries are made and will require new generations of scientists, ethicists, health care practitioners, policy makers, and patients to turn to individual cases to identify key issues and provide common language for discussion.

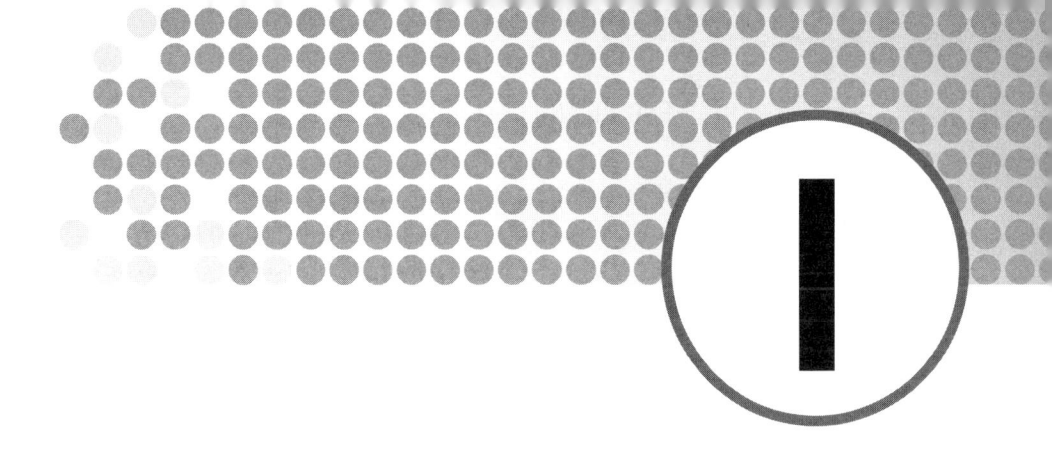

Ethical Principles in Medical Research

This chapter tracks the evolution of today's ethical standards for the use of human subjects in medical research, taking as its entry point the infamous Tuskegee *syphilis* study of the mid-20th century. Tuskegee is a key case in American medical ethics because of the harmful and deceptive nature of the research and because of the unjust selection of subjects, all of whom were poor and African-American. Revelations about the experiment in the early 1970s helped to reenergize *ethical principles* that had been codified after the Nazi atrocities of World War II and served as a driving force behind the seminal Belmont Report of 1979.

The chapter closes with a review of the *HIV vertical transmission* trials in low-income countries in the 1990s—a set of contemporary studies heralded by some physicians and ethicists as lifesaving and condemned by others as a "new Tuskegee." The vertical transmission trials reveal ethical complexities that have persisted long since the Tuskegee study, while also illustrating special issues that arise in the context of international research.

THE TUSKEGEE SYPHILIS STUDY

On the morning of July 26, 1972, the Associated Press (AP) broke a story nationwide that changed the face of American medical research. The story outlined a study conducted by the federal government in Tuskegee, Alabama, that used poor, mostly illiterate African-American men as "guinea pigs" to observe the effects of untreated syphilis.

The experiment was a "study in nature," designed to follow the natural course of the disease. Initiated by the U.S. Public Health Service (PHS) during the Great Depression, the Tuskegee study lasted 40 years and was the longest *nontherapeutic experiment* in medical history. Researchers followed approximately 400 African-American men with untreated syphilis and approximately 200 men without the disease, and when *penicillin* became standard treatment in the 1940s, the men were blocked from receiving it.

A Public Health Service researcher named Peter Buxton first raised concerns with government officials in 1966. Working as a venereal-disease investigator in the San Francisco office of PHS, he overheard talk of the study one day at lunch. "It didn't sound like what a PHS institution should be doing," he told James H. Jones, author of the book *Bad Blood: The Tuskegee Syphilis Experiment*. After reading published articles about Tuskegee, he wrote to Dr. William J. Brown, director of the PHS Division of Venereal Diseases, asking if the experimental intent was to gather information "on the syphilitic damage which these men were being allowed to endure," if the men had been told that they had untreated syphilis, and if they were being followed for autopsy. Over the next six years, Buxton's concerns were repeatedly dismissed and the study went forward. He brought the story to light in 1972, revealing details about it to a friend who was an AP reporter.

The Tuskegee study is considered a seminal case in the history of American medical research for several reasons: First, racist attitudes that permeated the mid-20th century United States allowed a study to proceed in an African-American population that would have been considered ethically unacceptable in a white population. Many physicians at the time believed unfair stereotypes—popular then among Caucasian Americans—that African Americans were unclean and promiscuous. One white physician typified such beliefs when he described African Americans as a "notoriously syphilis-soaked race."

Perhaps most telling, the study was never hidden from the rest of the medical profession; on the contrary, its results were widely reported in medical journals like the *Journal of the American Medical Association*. At least 17 articles about the study appeared in professional journals between the years of 1936 and 1972, yet not a single editor or doctor who encountered the study in these journals chose to blow the whistle publicly. One physician, Dr. Irwin J. Schatz, did write a letter to the Public Health Service questioning the morality of doctors involved in the research. "I am utterly astounded," Schatz wrote, "by the fact that physicians allow patients with a potentially fatal disease to remain untreated when effective therapy is available." He never received a reply.

Another defining feature of Tuskegee was that doctors clearly and purposefully deceived their patients. Though the Tuskegee study had grown out of a larger program to diagnose and treat African Americans for syphilis, when funding for the treatment program was pulled at the start of the Great Depression, researchers seized the opportunity for a long-term "study in nature." Patients' conditions were then hidden from them in order to prevent treatment and follow the course of the disease. The men were told that they had "bad blood" and that the *spinal taps* they received for diagnostic purposes were actually a treatment for their condition. The following letter from the Public Health Service to subjects in the study emphasized the importance of their cooperation in their "treatment":

Dear Sir:

Some time ago you were given a thorough examination and since that time we hope you have gotten a great deal of treatment for bad blood. You will now be given your last chance to get a second examination. This examination is a very special one and after it is finished you will be given a special treatment if it is believed you are in a condition to stand it.

The "special treatment" was no treatment at all, but a spinal tap—a diagnostic procedure that can be painful and carries a small risk of infection.

On July 27, 1972, the *New York Times* gave the following account of the way research subjects had been recruited: "They came around one day in 1932 and told Charles Pollard he could get a free physical examination the next afternoon at a nearby one-room school. 'So I went on over and they told me I had bad blood,' the 66-year-old farmer recalled today. 'And that's what they've been telling me ever since.'" According to the *Times* article, "Yesterday, Mr. Pollard learned that, for the last 40 years, he has had syphilis." In 1973, Pollard and another Tuskegee subject, Lester Scott, testified in Senate hearings that they thought "bad blood" meant something like low energy.

Dr. J. D. Williams, an African-American physician who interned at Tuskegee Institute Hospital and assisted with the experiment's *clinical* work, said, "The people who came in were

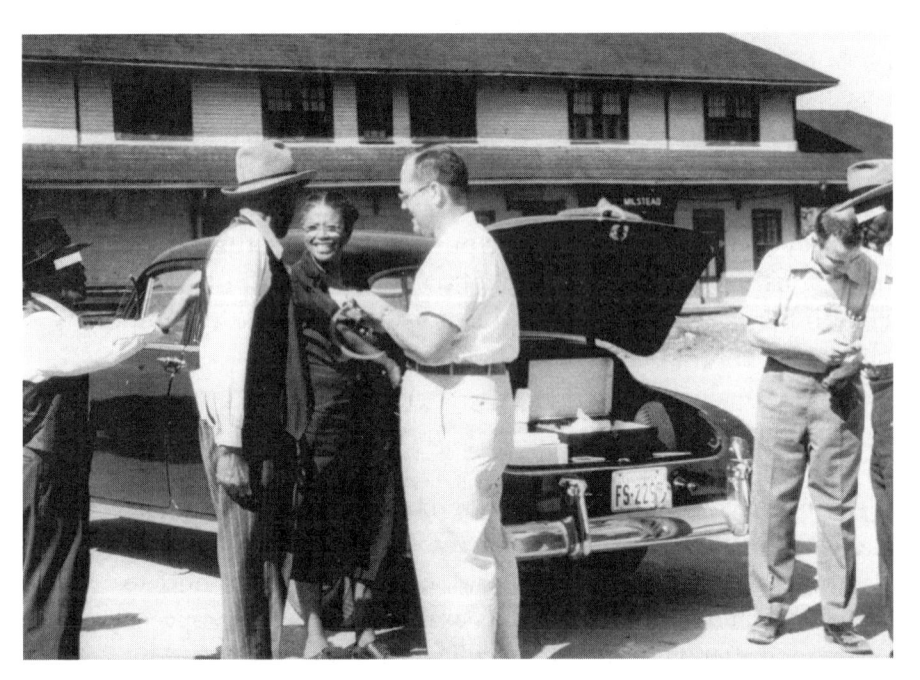

Subjects of the Tuskegee syphilis study meet with doctors and a nurse. The men were led to believe that they would receive treatment for "bad blood," when in reality they were subjects of an "experiment in nature" to observe the course of the untreated disease. *(Minority Health Archive, University of Pittsburgh, School of Health)*

not told what was being done. We told them we wanted to test them." He believed that subjects "thought they were being treated for rheumatism or bad stomachs," and was sure that the men were not told that doctors were looking for syphilis. "I don't think they would have known what that was."

By deceiving the men and their families, did researchers knowingly expose them to serious risks from untreated disease? Since the early 1900s, the heavy-metal compound *salvarsan*—called by its creator the "magic bullet"—had been somewhat successful in treating the symptoms of syphilis, but it was highly toxic and could cause severe side effects, even death. Penicillin later proved highly effective against the *spirochete* (corkscrew shaped bacterium) that causes syphilis, and the drug became widely available in the 1940s. None of the men recruited at Tuskegee received penicillin through the study. Even as the drug became standard treatment, researchers blocked the men's participation in free public treatment programs, and young men were even exempted from the draft in order to prevent their treatment by military doctors.

Apologists for the study have argued that the men, who lived in a poor rural area of Alabama, probably would not have received diagnosis or medicine anyway, and that it would be impossible to prove that they were harmed by lack of treatment. Penicillin's beneficial effects on early stages of the disease were well-documented by the 1940s, but its effects on later stages of the disease (which the men in the Tuskegee study were suffering from by that point) were less certain. The men, of course, were never informed of their medical options or given the opportunity to decide for themselves whether or not to be treated with the best available medicine.

Spouses were never told of their loved ones' conditions, and no attempt was made to survey or treat wives or children, even though 62 percent of patients admitted to the Tuskegee program had *congenital* syphilis (syphilis present at birth). Any number of family members may have contracted syphilis as a result of the researchers' deception. It is now known that the disease is rarely contagious in its later stages, but when the study began, doctors did not know this to be the case. Either they neglected to consider the risks to families, or they judged potential risks to be less important than the goals of the study.

In addition to clear ethical concerns, critics pointed to the poor scientific value of the study's results. Dr. James B. Lucas, assistant chief of the PHS venereal disease branch in 1970, called the experiment "bad science" and said, "Nothing learned will prevent, find, or cure a single case of infectious syphilis or bring us closer to our basic mission of controlling venereal disease in the United States."

Experiments of this scope usually employ a formal scientific *protocol*—a standard document laying out research procedures in a step-by-step fashion. Such a document never existed for Tuskegee. In the early years of the study, a *control* (comparison) *group* of men being treated for syphilis was dropped—leaving a group of men who did not have syphilis at all to act as "controls"—so that nothing new was learned about the course of the treated versus the untreated disease. Recordkeeping procedures were poor, and names of men living with the disease were often confused with men free from the disease. Several "controls" eventually contracted the disease and were moved to the "subject" group. Ultimately, many of the subjects did find some treatment with doctors outside of the study, casting doubt on the study's findings about the course of the untreated disease.

Tuskegee left deep and permanent marks on the practice of medical research in the United States. In the wake of the scandal, the National Commission for the Protection of Human Subjects of Biomedical and Behavioral Research was created, and in April 1979 that commission released its landmark Belmont Report—so named for the Belmont Conference Center just south of Baltimore where it was drafted. The commission's findings are outlined in the next section, as are three other ethical cornerstones of U.S. medical research at home and abroad: the Nuremberg Code, the Declaration of Helsinki, and federal regulations on the protection of human subjects.

Despite improved protections, sociological studies have shown that Tuskegee, and the attitudes that tolerated it, have produced rampant and persistent mistrust of medical and public health authorities among African Americans. Chapter 7 will

Ethical Principles in Medical Research 7

Tuskegee survivor Herman Shaw, 94, receives an official apology from President Clinton in May 1997. *(AP Photo/Greg Gibson)*

revisit this issue, and the role it may play in lower life expectancies among African Americans even today.

On May 16, 1997, five of the eight remaining survivors of the Tuskegee study attended a White House ceremony, where they received a formal apology from President Bill Clinton. "What was done cannot be undone, but we can end the silence," the president said. "We can stop turning our heads away. We can look at you in the eye, and finally say, on behalf of the American people, what the United States government did was shameful and I am sorry."

ETHICAL PRINCIPLES: NUREMBERG TO BELMONT

Egregious abuses like those committed in Tuskegee seem far removed from most young Americans' experience, yet the study ended just a few decades ago. What has changed in the meantime? To what ethical codes are health care professionals now held accountable when treating patients and performing research, and do all research subjects enjoy those protections equally?

The following is a brief overview of four sets of ethical guidelines governing medical research, laid down since the Nazi atrocities of World War II. Together they form the evaluative framework for the treatment of human subjects, though researchers and ethicists often disagree on their implications for contemporary research practices.

The Nuremberg Code

Revelations about barbaric experiments performed before and during World War II by Nazi doctors on thousands of men, women, and children imprisoned in concentration camps led to a criminal trial known as the "Doctors' Trial," which began on December 9, 1946. Karl Brandt and 22 other doctors and Nazi officials were tried before an American military tribunal at Nuremberg, Germany, for their participation in these experiments. For almost 140 days, the tribunal listened to testimony from 85 witnesses and reviewed almost 1,500 documents detailing cruel procedures in which thousands of mostly Jewish, Polish, Russian, and Roma (Gypsy) victims died or were

Ethical Principles in Medical Research

left permanently disabled. On August 20, 1947, the American judges convicted 16 of the defendants of "war crimes and crimes against humanity." Seven were sentenced to death, and they were executed on June 2, 1948.

Along with the verdict, the judges delivered their opinion on what should be the ethical requirements for medical experimentation on human beings. This opinion, subsequently known as the Nuremberg Code, formed the bedrock of all future U.S. and international codes of ethics for medical research on human subjects. It codified the notion of "voluntary consent," requiring that a potential subject first be told the nature and risks of the research and then be given the opportunity to "exercise free power of choice" with no coercion. Other central requirements of the code focused on the protection of subjects from "all

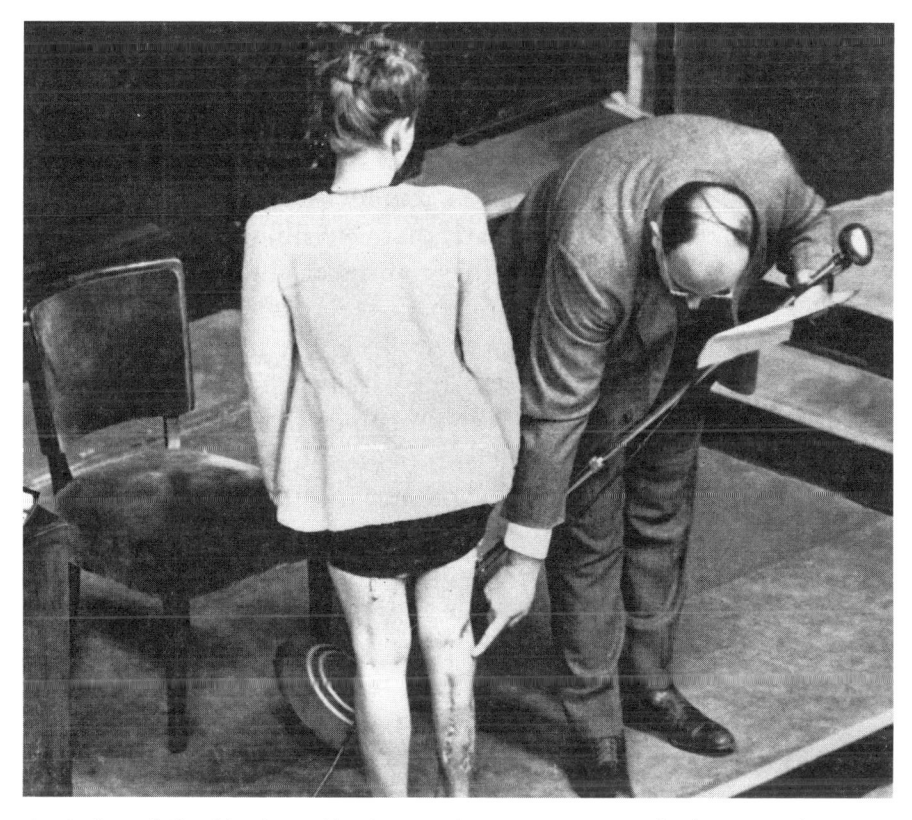

A victim of the Nazi medical experiments appears before a military tribunal at the Nuremberg "Doctors' Trial." *(Granger)*

unnecessary physical and mental suffering and injury," rigorous scientific design, and the freedom of subjects to stop participating at any point in the experiment.

The Nuremberg Code was delivered in 1947, 15 years after Tuskegee began. Yet even with the code's explicit consent requirements, for another 25 years Tuskegee researchers deceived their patients. Why? The code was disseminated among government officials, but it was not aggressively enforced with medical practitioners until much later. The tradition of paternalism, under which it was common for patients simply to trust their doctors and for doctors not to obtain consent from sick patients (as opposed to *healthy volunteers,* for whom it was standard practice to obtain consent), persisted into the 1970s.

The Declaration of Helsinki

Following World War II and the Nuremberg Doctors' Trial, the Medical Ethics Committee of the World Medical Association (WMA) embarked on an unprecedented project: The creation of an international code of ethics to guide doctors all over the world in their research with human subjects. After more than a decade of research, discussion, and revision, the WMA General Assembly adopted the Declaration of Helsinki in June 1964. The Declaration went even further to protect human subjects than the Nuremberg Code, stating that "[c]oncern for the interests of the subject must always prevail over the interests of science and society." It stressed that any patient "should be assured of the best proven diagnostic and therapeutic method."

The Declaration of Helsinki is a landmark in the history of medical ethics as it was the first unified effort on the part of the international medical community to self-regulate. It has played a key role in contemporary debates over medical research conducted in developing nations (discussed later in this chapter).

U.S. Regulations on the Protection of Human Subjects

The Nuremberg Code and the Declaration of Helsinki set the stage for U.S. regulations governing federally funded research with human subjects. Disseminated by the Department of Health and Human Services, these regulations include requirements for *informed consent* and lay down the structure and responsibilities of "institutional review boards," or *IRBs*—local committees charged with the review and evaluation of medical research at hospitals, clinics, universities, and other research organizations (as detailed in the sidebar below). All potential research involving human subjects must first be scrutinized by an IRB, which is supplied with specific standards by which to judge the scientific and ethical merits of experiments.

The Belmont Report

In 1979, the National Commission for the Protection of Human Subjects of Biomedical and Behavioral Research published its Belmont Report, which laid out three basic ethical principles:

1. *Respect for persons.* The commission recognized the importance of preserving personal *autonomy,* or the ability of a person to make independent choices, primarily through guidelines for informed consent. "Respect for persons," the commission wrote, "requires that subjects, to the degree that they are capable, be given the

(continues on page 14)

INSTITUTIONAL REVIEW BOARDS

First promulgated in 1974, federal regulations on the protection of human subjects require that a university or other research institution seeking or receiving federal funds maintain an Institutional Review Board (IRB)—a committee of at least five members charged with reviewing the scientific and ethical merits of research on human subjects. Many institutions use IRBs to review all such research projects, not just those receiving federal funds.

(continues)

(continued)

Ethical Principle	Respect for Persons	Beneficence	Justice
Regulatory review requirements	Informed consent is obtained from research subjects or from their legally authorized representative(s).	The proposed research design is scientifically sound and will not unnecessarily expose subjects to risk. Risks to subjects are reasonable in relation to anticipated benefits of the research. Risks to subjects are minimized. Subject privacy and confidentiality are maximized.	Subject selection is equitable. Additional safeguards are in place for subjects likely to be vulnerable to coercion or undue influence.
Suggested questions for IRB discussion	Does the informed consent document include the required elements? Is the consent document understandable to subjects? Who will obtain informed consent (primary investigator, nurse, or other) and in what setting? If appropriate, is there a children's assent? Is the IRB requested to waive or alter any informed consent requirement?	Will the research yield generalizable knowledge, and is it worth exposing subjects to risk? Is there prospect of direct benefit to subjects? Does the research design minimize risks to subjects? Will identifiable personal research data be protected from access and use to the extent possible? Are any special privacy and confidentiality issues properly addressed, e.g., use of genetic information?	Who is to be enrolled (men, women, ethnic minorities, children, seriously ill persons, healthy volunteers)? Are these subjects appropriate for the protocol? Are appropriate protections in place for vulnerable subjects, e.g., pregnant women, fetuses, or the socially or economically disadvantaged?

© Infobase Publishing

According to the regulations, each committee must include "one member whose primary concerns are in scientific areas and at least one member whose primary concerns are in non-scientific areas." An IRB should have the collective professional expertise to perform thorough scientific, ethical, and legal analysis of research. The institution should endeavor to provide appropriate gender representation, and if the IRB commonly reviews research proposals involving *vulnerable populations* (see chapter 2), then "consideration shall be given to the inclusion of one or more individuals who are knowledgeable about and experienced in working with these subjects." Each IRB must contain at least one member who is not affiliated with the institution, and no IRB may consist solely of members of one profession.

Before a researcher receives federal money, the written experimental protocol and informed-consent forms must be reviewed by the IRB in order to weigh the risks and benefits of the research, to ensure that informed-consent guidelines are satisfied, and to protect vulnerable populations from possible exploitation. IRBs may approve or reject research proposals, or they may require researchers to change certain aspects of experiments before issuing approval.

IRBs are administered by the Office for Human Research Protections (OHRP), a branch of the Department of Health and Human Services charged with protecting human subjects and investigating allegations of unethical research practices.

(opposite) Three central ethical principles and their influence on the IRB review process. *(Regulatory requirements and sample review questions adapted and abridged from the NIH Office of Human Subjects Research, "IRB Protocol Review Standards,"* Guidelines for the Conduct of Research Involving Human Subjects at the National Institutes of Health, *April 2004)*

(continued from page 11)

>   opportunity to choose what shall or shall not happen to them. This opportunity is provided when adequate standards for informed consent are satisfied." Though exact requirements for informed consent are debatable and difficult to standardize, the commission broke the consent process down into three basic elements: information, comprehension, and voluntariness.
> 2. *Beneficence.* Fulfilling this principle means both minimizing possible risks and maximizing possible benefits to research subjects. The commission recognized the challenge in "making precise judgments" about risks and benefits when those risks and benefits cannot always be known, but stressed that "the idea of systematic, nonarbitrary analysis of risks and benefits should be emulated insofar as possible." In other words, researchers should do their absolute best to ensure that their subjects' interests are served.
> 3. *Justice.* The equal or fair distribution of the burdens and benefits of research was also at issue. The commission was troubled that in the past, "some classes" of people—for example, minorities, prisoners, and the economically disadvantaged—had been "systematically selected" for research "simply because of their easy availability, their compromised position, or their manipulability," and it recommended that publicly funded research "not provide advantages only to those who can afford them" and "not unduly involve persons from groups unlikely to be among the beneficiaries of subsequent applications of the research."

The next section looks at how these ethical principles are applied in medical research today. Do all people enjoy equal protections? To what degree do economic inequalities, racial stereotypes, and other potential sources of discrimination figure into the design and execution of medical research? The HIV vertical transmission trials conducted in developing countries in the 1990s exposed some important limitations of prevailing ethical standards.

THE HIV VERTICAL TRANSMISSION TRIALS

Among the most controversial medical experiments in recent years, the HIV drug trials of so-called vertical (mother-to-child) transmission put ethical codes of conduct to the test. They proved that even the most seemingly straightforward principles can be extraordinarily complicated to apply in practice. Condemned by some physicians and medical ethicists as a "new Tuskegee," these drug trials were heralded by others as a valiant and necessary attempt to save millions of lives.

Special issues arise in international medical research, particularly when studies are conducted in relatively resource-poor countries where the *standard of care* is lower than that of sponsoring countries. Of central concern is that the poor quality of health care in parts of the developing world might be used to justify research that would not meet ethical standards in the researchers' home country.

The HIV vertical transmission trials were conducted in five African nations, the Dominican Republic, and Thailand, and involved 12,211 pregnant women. Very few generalizations can be made about these studies since so many of them differed in their implementation, though some common issues did arise that may impact future international research (such as HIV vaccine trials).

First, the tragic backdrop: By the mid-1990s, the HIV/AIDS epidemic was raging throughout many parts of the developing world. An estimated 1,000 babies were born every day with HIV worldwide. The chances that these infants or their mothers would receive life-extending drugs were extraordinarily slim. At the time, drug combinations cost approximately $15,000 per person annually in developing countries—a price tag well beyond the reach of most people living in sub-Saharan Africa, where annual health care expenditures were in the range of $10 per person and where the screening procedure for HIV alone costs that amount. In countries like Malawi, South Africa, and Uganda, as many as 40 percent of pregnant women were infected and roughly one in

AIDS orphans with their grandmother in Mombasa, Kenya. Three of the children are HIV positive. *(©Sean Sprague/The Image Works)*

four or five of their babies would be born HIV-positive. If infants were not infected at birth, they stood a significant chance (14 percent) of being infected through breastfeeding. In many of these countries, nutritious food and clean water were in short supply, and the risk of starvation and dehydration outweighed the risk of HIV transmission. Breast-feeding, therefore, was still recommended to HIV-positive mothers.

In the United States and other industrialized nations, the outlook for infants born to HIV-positive mothers was markedly different. In 1994, U.S. and French researchers had announced the remarkable results of a study administering a regimen of the drug *AZT* (now called *ZDV*), first to pregnant mothers and then to their infants after birth. The new regimen—known as the 076 regimen (its *clinical trial* number)—reduced mother-to-infant transmission by two-thirds. By June 1994, the U.S. Public Health Service had recommended that the 076 regimen be given

Ethical Principles in Medical Research

to HIV-positive pregnant women as standard treatment. But would this new breakthrough treatment reach pregnant women in developing nations? Unlikely, since it still was prohibitively expensive ($800 to $1,000 for the AZT alone) and since most AIDS-stricken African nations did not have the medical facilities or staff to administer such a complex drug regimen.

It was this gaping disparity in infant care that the World Health Organization, the United Nations, the National Institutes of Health (NIH), and the Centers for Disease Control and Prevention (CDC) tried to address in Geneva in June 1994 when they came together to design studies of whether a shorter course of AZT would help prevent mother-to-infant transmission. "In the absence of any standard of care," reported the *New York Times* in September 1997, health officials "settled on comparing different courses of AZT to dummy medications"—in other words, they decided to use a comparison (control) group made up of people who received only a *placebo*, or fake drug. Almost no one disputed the need to find cheaper and easier-to-administer regimens for developing countries. The use of placebos, however, was (and still is) a source of heated international controversy.

After less than four years of research, the CDC released results of a study in Thailand that showed that a shortened course of AZT reduced the mother-to-infant transmission rate by 50 percent over the placebo group. This course of treatment would cost $80, or approximately one-tenth the price of the 076 regimen. The next year, a large placebo-controlled study based in three African nations demonstrated that an even shorter course of treatment reduced the transmission rate by 37 percent. Despite these results, hailed by United Nations AIDS official Joseph Saba as ensuring that "we can save a lot of children using several strategies, whichever makes sense for the women, their doctors or the community in which they live," there was bitter controversy over the cost of the research in human terms. Major concerns with the vertical transmission studies can be grouped under two ethical principles expressed in the Belmont Report: respect for persons and beneficence.

Respect for Persons: Informed Consent in a Cross-Cultural Setting

Obtaining meaningful informed consent presents challenges in any research setting, even when the researcher and research subject speak the same language. Where language barriers are magnified, so are the ethical challenges. In one South African trial in 1997, the word *placebo* was translated sometimes as being a "chuff-chuff" drug or "pretend" drug, and at other times as being a "spaza" drug, generally understood to be "half the real thing." This second translation would have misled participants, since it would have caused them to believe that they might receive helpful medicine—even if only half as helpful—when in fact they received no medicine at all.

It appears that some subjects in the HIV vertical transmission trials understood little of what they were consenting to or what the risks might be. One participant in a trial conducted in Abidjan, Ivory Coast, told the *New York Times* in October 1997, "They gave me a bunch of pills to take, and told me how to take them. Some were for *malaria,* some were for fevers, and some were supposed to be for the virus. I knew that there were different kinds, but I figured that if one of them didn't work against AIDS, then one of the other ones would." Another woman told the same reporter, "They said that it would help my child, and that it would ease my childbirth too."

Lack of education of participants has often been cited as a problem, yet other factors must have been at play. One educated woman in the Ivory Coast, a 31-year-old mother with a law degree, said that it had never been explained to her that AZT was a proven treatment elsewhere in the world for reducing mother-to-child transmission. When asked how she would feel if she learned that she had been in the placebo group, she replied, "I would say quite simply that that is an injustice."

It is unclear why some mothers did not understand that they might receive a placebo while others did. "People are trying to help us," said one woman, "and if a bunch of people have to die first, I am ready to risk my life too, so that other women and their babies can survive. If I got the placebo, that will hurt, for sure. But there is no evil involved."

Respect for Persons: The Problem of Coercion

Where people are extremely poor and where benefits of becoming research subjects are offered, such as payment for transportation to the research site and meals for the day, can consent be considered voluntary? South African medical ethicist Keymanthri Moodley says not. "Attractive offers such as free medication or extra money can leave persons without any meaningful choice apart from accepting the offer largely because such persons are constrained in a desperate situation. Whatever we may decide to call this, it is widely held that offers of this magnitude to a person in desperate need is inherently exploitative and is not consistent with autonomous choice." Moodley points out that in South Africa, it is standard research practice to pay drug-trial participants six dollars per visit, and that it is problematic to include this amount in the informed consent document since people from poor communities will consent for half that amount. One woman who agreed to take part in the HIV vertical transmission trial in the Ivory Coast was asked why she did, and she answered, "The medical care that they are promising me."

Factors other than money or free medical care might cause a patient to feel coerced. Her relationship to authority may affect her participation; even if she has reservations, mistrust or even fear may persuade her to consent. Research conducted in Durban, South Africa, on the question of whether consent for HIV testing could be considered informed or voluntary found that in a majority of cases it could not. Of the 56 women interviewed, 88 percent felt compelled to participate in HIV testing, despite assurances to them by researchers that their choice would be perfectly voluntary. Twenty-eight percent believed that if they chose not to participate, their medical care would somehow be compromised.

Beneficence: Defining "Standard of Care"

Most of the controversy surrounding the use of placebos hinged on the question: "To what standard of care should new treatments be compared?" Critics of the mother-to-child transmission studies asserted that similar trials never would have been permitted in the United States, since patients were not being given the best current treatment. Doctors Peter Lurie and Sidney Wolfe of the consumer-advocacy group Public Citizen wrote to Secretary of Health and Human Services Donna Shalala that "as many as 1,002 newborn infants in Africa, Asia, and the Caribbean will die from unnecessary HIV infections" contracted in vertical transmission trials funded by the NIH and the CDC. They pointed to two trials of treatments to reduce HIV transmission conducted in the United States, in which all of the participants—including those in the control group—were given AZT or other antiretroviral drugs.

Tests involving placebos are rarely used in the United States, particularly when a potentially deadly disease like HIV is involved. In recent years it has become common to conduct *active control trials,* in which a control group receives the standard treatment for a condition (rather than a placebo) and a treatment group receives the experimental treatment. In this case, since the full 076 regimen had been proven quite effective in the industrialized world, critics argued that it should have been administered to a control group instead of placebos.

Dr. Marcia Angell, executive editor of the *New England Journal of Medicine,* argued in a September 1997 opinion that the concept of "standard of care" should always be interpreted in universal rather than local terms. The Declaration of Helsinki, she pointed out, called for the "best proven diagnostic and therapeutic method" for research subjects—including those in the control group—and federal regulations on the protection of human subjects required that U.S. researchers provide protections for foreign subjects at least equivalent to those existing in the United States. (Since the controversy over the vertical transmission trials, the Declaration of Helsinki has been revised to allow for the use of placebos, but only in cases where there is an "absence of existing proven therapy" or where placebo use is "necessary to determine the efficacy or safety" of a diagnostic method or treatment.)

Justifications for the HIV vertical transmission trials, Angell wrote, are "reminiscent of those for the Tuskegee study: Women in the Third World would not receive antiretroviral treatment anyway, so the investigators are simply observing what would happen to the subjects' infants if there were no study." Where clinical trials have become "big business," she wrote—where they are governed by pressures to get the work done as quickly and efficiently as possible—"it seems as if we have not come very far from Tuskegee after all."

Proponents of the studies called the comparison to Tuskegee inaccurate and offensive. When asked in November 1997 if the analogy had merit, Jack Killen, director of the National Institute of Allergy and Infectious Diseases' Division of AIDS said, "Absolutely not. The Tuskegee analogy is inappropriate because the men in that study were prevented from receiving treatment that was widely available to others in the United States, and they were deliberately misled. Women enrolled in the perinatal transmission studies sponsored by NIAID all receive the level of care they would receive if not in the trial, and the complete study design, including the fact that they may receive a placebo, is fully explained to them."

Only placebos, proponents argued, would realistically simulate the standard of care in host countries. The head of the CDC's AIDS program, Helene Gayle, held that "Part of doing ethical trials is that you are answering questions that are relevant for those countries." And the most relevant question, said Killen, for "HIV-infected women, their infants, and those who care for them in developing countries" was whether the new treatment improved upon what was currently available—i.e., no antiretroviral treatment at all.

Implicit in criticisms of the trials was a belief that the principle of beneficence, which seeks to maximize benefits and minimize harm, had been violated by the use of a placebo. "Investigators are responsible for all subjects enrolled in a trial," Angell wrote, "not just some of them, and the goals of the research are always secondary to the well-being of the participants." Advocates of the studies also appealed to the principle of beneficence as laid out in the Belmont Report, since it emphasized not just the well-being of research subjects but also the obligation of investigators "to give forethought to the maximization of benefits and the reduction of risk that might occur from the research investigation."

SUMMARY

The authors of the Belmont Report predicted conflicts about what it means to minimize harm and maximize benefit when they wrote, "with all hard cases, the different claims covered by the principle of beneficence may come into conflict and force difficult choices." The HIV vertical transmission trials have been over for several years, yet the ethical dispute over their use of placebos persists. Similar issues are expected to arise with HIV vaccine tests in the coming years.

The vertical transmission trials also reveal many of the complexities involved in obtaining truly informed and voluntary consent—a process that gets more challenging when technology is more complex and subjects are more vulnerable. The authors of the Belmont Report acknowledged problems inherent in the informed-consent process, noting that "While the importance of informed consent is unquestioned, controversy prevails over the nature and possibility of informed consent," and that "it is impossible to state precisely where justifiable persuasion ends and undue influence begins." The application of ethical principles to real-life research is anything but simple. Thorough IRB review and open, informed debate are integral to tailoring broad ethical principles to meet the challenges presented by complex cases.

The high-speed chase of medical discovery may well require more than mere refinements of the ethical framework by which research is judged. The relatively new fields of *genetic engineering* and *xenotransplantation,* for example, pose risks not just to consenting research subjects but also to the population at large (see chapter 3), and may require an entirely new set of ethical guidelines.

2

Vulnerable Populations

The last chapter highlighted the Tuskegee syphilis study, in which doctors and researchers clearly violated ethical principles in their use of patients in medical research. Today there are strong ethical codes in place, making transgressions like Tuskegee seem shocking and far removed from most Americans' experience. Yet the stunning pace of medical advancement requires vigilance from patients, families, health care practitioners, ethicists, and policy makers to ensure that the laudable ends of finding new cures and treatments not be used to justify ethically suspect research. This chapter takes a closer look at issues with vulnerable populations—groups of research subjects who may require extra protection.

A major focus of the modern medical ethics movement since Tuskegee has been protecting society's most vulnerable members—and citizens of other countries—from harm and deception and to ensure that they receive the best possible care. Senator Ted Kennedy framed the problem during congressional hearings on human experimentation in 1973: "Those who have borne the principal brunt of research—whether it is drugs or even experimental surgery—have been the more disadvantaged people within our society; have been the institutionalized, the poor, and minority members."

Most medical ethicists consider the following groups to be vulnerable: Children, economically and educationally disadvantaged people, racial and ethnic minorities, prisoners and institutionalized people, mentally or emotionally disabled people, and terminally ill or critically ill people. Some groups are considered vulnerable because they are dependent upon others to make decisions in their best interests (children, mentally disabled adults, critically ill people). Others may be too easily coerced by money or force (economically disadvantaged people and prisoners), and others may suffer from long-standing effects of social prejudice (African Americans and other racial and ethnic minorities) or language and cultural barriers (recent immigrants and citizens of other countries). Sometimes more than one factor is at play.

The use of a vulnerable population can be so clearly wrong that when the public becomes aware of the research in question, it is stopped. Experiments on inmates at Philadelphia's Holmesburg prison, for example, were halted in 1974 after it was revealed that inmates had received chemical burns and had been administered hazardous—even potentially deadly—substances like radioactive isotopes, chemical warfare agents, and dioxin (a potent carcinogen, or *cancer*-causing agent). One inmate, Leodus Jones, told Senator Kennedy at the 1973 hearings on human experimentation that "The only way I seen that I was able to raise bail money . . . was to submit to medical testing." Public outcry about the use of captive or institutionalized populations in medical research brought much stricter controls on the use of prisoners and other captive populations in the 1970s.

Other cases involving vulnerable populations are not so straightforward, since the line between harming and healing is often blurred or as yet undiscovered (as in the testing of a new *chemotherapy* drug, for example); since communication, language, and cognitive barriers can make informed consent imperfect or impossible; and since what makes a population vulnerable sometimes goes hand-in-hand with important reasons to involve them in research in the first place (see the discussion of research with children below). The drive to find new treatments and cures can be compelling for researchers, who may then use these worthy ends to justify ethically troubling research and patient care. Moreover, patients and healthy research subjects (or their guardians) are not always as proactive as they might be with decisions about medical care because of respect for authority, because they wish not to appear ignorant or to ask "stupid questions," or because they trust that their interests will always come first.

This chapter details past and present examples of questionable research involving two vulnerable groups—children and emergency-room patients—and considers special issues researchers must consider when working with them. In some cases outlined here, the subjects can be considered vulnerable in more than one way. At Willowbrook State School, for example, children recruited for research were also mentally retarded and institutionalized.

RESEARCH WITH CHILDREN: THE WILLOWBROOK HEPATITIS STUDY

In the late 1960s, the American media caught on to an experiment conducted at Willowbrook State School in Staten Island, New York, a long-term care institution for severely mentally retarded children. Since 1956, Saul Krugman and Joan P. Giles of the New York University School of Medicine had led a team of researchers in a study of the natural course of viral *hepatitis* (inflammation of the liver) and of the possible effectiveness of *gamma globulin* (a substance extracted from the blood) as an inoculation against the disease.

Krugman and Giles chose Willowbrook because viral hepatitis was *endemic* to the institution, meaning that it was always present and children living there very commonly contracted it. According to the researchers, under conditions of chronic exposure, "most newly admitted children became infected within the first six to 12 months of residence in the institution." Hepatitis is typically mild, but sometimes a patient's liver can be permanently damaged from it.

Unlike the Tuskegee study, the Willowbrook study is considered to have contributed important information to the body of medical knowledge. Moreover, the researchers did obtain consent from the parents of children involved in the study and did not use children for whom parental consent was not given.

What most surprised critics at the time was the researchers' choice to begin deliberately infecting children with a mild strain of hepatitis in order to ensure that the best data could be collected before, during, and after the moment of infection.

In 1966, Harvard anesthesiologist Henry K. Beecher wrote "Ethics and Clinical Research," a landmark piece published in that year's *New England Journal of Medicine.* The article highlighted Willowbrook and 21 other ethically troubling experiments Beecher had found in contemporary medical journals. "Evidence is at hand," Beecher began his account, "that many of the patients in the examples to follow never had the risk satisfactorily explained to them, and it seems obvious that further hundreds have not known that they were the subjects of an experiment although grave consequences have been suffered as a direct result of experiments described here." He found it troubling that parents of children recruited at Willowbrook "gave consent for the intramuscular injection or oral administration of the virus, but nothing is said regarding what was told them concerning the appreciable hazards involved . . . There is no right to risk an injury to one person for the benefit of others."

The researchers defended deliberate infection of children with a mild strain of the virus as being in the subjects' best interests; susceptible children, they argued, would become infected anyway and under more dangerous conditions. The children involved in the study would receive special care and protection from other diseases common at Willowbrook, and infection with the milder form of hepatitis should have a protective effect against more harmful forms. "It should be emphasized," the researchers said, "that the artificial induction of hepatitis implies a 'therapeutic' effect because of the immunity which is conferred."

Critics of the study remained unconvinced, and some highlighted the vulnerability of children at Willowbrook as a serious issue. Dr. Stephen Goldby called the study "quite unjustifiable," and asked in a letter to the British medical journal *Lancet,* "Is it right to perform an experiment on a normal or mentally retarded child when no benefit can result to that individual? I think that the answer is no, and that the question of parental consent is irrelevant."

Others focused on issues of informed consent and on the risks to subjects who might never have contracted the disease outside of the study. There were questions about the wording of the consent form, which may have deceived parents into thinking that their children were receiving a vaccine against the virus, and also about possible coercion, since the waiting list to get into the overcrowded institution was long and some parents were told that there was room only in the "hepatitis unit." In another letter to the *Lancet,* Dr. M. H. Pappworth suggested that the researchers only claimed *"therapeutic* effects" after the fact to justify the research, and that protection of the child subjects was never an intended purpose. "No doctor," wrote Pappworth, "is ever justified in placing society or science first and his obligation to patients second. Any claim to act for the good of society should be regarded with distaste because it may be merely a highflown expression to cloak outrageous acts." On this side of the Atlantic, however, the editors of such prestigious journals as the *New England Journal of Medicine* and the *Journal of the American Medical Association* defended the research.

The Willowbrook study, like the Tuskegee study, was never kept secret. Quite the opposite: Several official review bodies approved the research, including the New York Department of Health, the executive faculty of the New York University School of Medicine, and the Armed Forces Epidemiological Board (which funded the research). Nonetheless, the *New York Times* and *Wall Street Journal* both ran stories about the study on March 24, 1971, and it became a heated topic of debate in the medical literature in the early 1970s. The study was finally terminated in 1972.

PROTECTIONS FOR CHILDREN

Today there are guidelines protecting children in medical research funded or supported by the federal government, thanks in part to public concern over studies like Willowbrook. Most current research with children must present "no greater than minimal risk to the children," where *minimal risk* means that "the probability and magnitude of harm or discomfort anticipated in the research are not greater in and of themselves than those ordinarily encountered in daily life or during the performance of routine physical or psychological examinations or tests." If research presents more than minimal risk, it should either hold the "prospect of direct benefit" to the child or else subject the child to only a minor increase in risk. But why use children in research at all? Why—if society believes that children should not come to harm if at all possible—expose them to even the smallest of risks?

In medical science, children cannot be treated like miniature adults. Their bodies and minds are growing and changing. Their anatomy, biochemistry, and metabolic systems are different from adults'. Drugs often affect children differently than they do adults, and special surgical techniques must be devised that work with children's physiology. Certain problems seen in children are never seen in adults (congenital heart defects, for example, that must be corrected in order for children to survive). Says medical ethicist Ronald Munson, "For many medical purposes, children must be thought of almost as if they were wholly different organisms. Their special biological features set them apart and mark them as subjects requiring special study. To gain the kind of knowledge and understanding required for effective medical treatment of children, it is often impossible to limit research solely to adults."

Yet the biological differences that make children important to study can also make them physically and emotionally vulnerable. Researchers must take extra care to protect their growing, changing organs and systems. Children over the age of seven typically must give their "assent" to be involved in a research study, but it is their parents or guardians who ultimately decide whether participation is in the child's best interests.

A source of continuing controversy is when, if ever, the choice to participate in medical research should lie with the child. Some teenagers, for example, are better able to understand possible benefits and hazards of a medical study than are some adults. Another question frequently considered in the medical ethics literature since Willowbrook is whether children should ever be involved in research that does not offer them direct therapeutic benefit; the Maryland judges who faulted the Kennedy Krieger Institute for a lead-paint study in children thought not.

THE KENNEDY KRIEGER INSTITUTE LEAD-PAINT STUDY

On August 16, 2001, Maryland's highest court denounced a lead-paint study in children conducted in low-income Baltimore neighborhoods by the Kennedy Krieger Institute (KKI), a children's hospital and research affiliate of the Johns Hopkins University Medical School. The Institute had played an important historical role in the fight against *lead poisoning* in children, and in 1993 received a $200,000 grant from the U.S. Environmental Protection Agency (EPA) to test how well different levels of *lead abatement* (reduction of lead hazards) in East Baltimore rental housing would reduce lead levels in the blood of inner-city children. KKI helped landlords find public funding for repairs and encouraged them to rent the premises to families with young children. Families with children already living in the homes were encouraged to stay.

Lead remains one of the primary environmental hazards for children in the United States. Leaded gasoline caused severe brain damage to thousands of children in the 1970s, but now gasoline is "unleaded," and the primary route of exposure is ingestion (swallowing) of lead-based paint. More than one in 25 American children have blood lead levels high enough to reduce IQ or cause learning disabilities, hyperactivity, or violent behavior. Extremely toxic levels of lead can cause seizures, *coma*, even death.

Young children are especially susceptible to lead poisoning because their hands are frequently in their mouths, because their bodies can absorb 40 to 50 percent of the lead that they ingest (adults absorb 10 percent), and because even small amounts of lead can cause dangerously high concentrations in small bodies. Children's growing, developing brains are particularly vulnerable to lead's damaging effects.

Many homes built before 1978 contain at least some lead-based paint, and contamination can be especially severe in

A volunteer dry-scraping lead-based paint from a porch railing with children nearby. Dry-scraping without proper protective gear and cleanup can produce a hazardous level of lead exposure, especially for young children. *(CDC/Aaron L. Sussell)*

low-income neighborhoods where poor maintenance of older homes can increase the chances that children will swallow lead dust and paint chips. In the early 1990s when the KKI study commenced, an estimated 35 to 40 percent of children from low-income urban neighborhoods throughout the country had hazardous levels of lead in their blood (in sharp contrast to only 5 percent of non-Hispanic white children from outside urban areas). In some older cities like Baltimore, this percentage was thought to be even higher. KKI estimated that up to 95 percent of low-income housing units in inner-city Baltimore neighborhoods were lead-contaminated, and that as many as two-thirds of children living in these neighborhoods had elevated levels of lead in their blood.

The study recruited 108 families with healthy children into five different test groups: 25 families lived in homes with minimal repairs, including scraping lead-based paint; 25 lived in homes with an intermediate level of repair; 25 lived in homes with extensive work, including replacement of old windows (a major source of lead dust); and the rest were divided between two control groups in homes where all lead hazards had been removed or in newer homes that had never contained lead.

Two mothers of children enrolled in the study filed lawsuits against KKI, claiming that the Institute never warned them of the potential hazards of lead poisoning. The lawyer for one of the mothers told the *New York Times* that when her client moved into rental housing as part of the experiment in May 1994, her son's blood lead level was safe and she was unaware of the study. "After she moved in," the lawyer contended, "Kennedy Krieger enrolled her in the study, and she signed the informed consent, but no one ever told her, 'There's lead in this house, and it can cause brain damage.'" A month after moving into the new home, the child allegedly had elevated blood lead levels.

Lower courts dismissed the cases, but on August 16, 2001, the Maryland Court of Appeals overturned those decisions and allowed the mothers to sue. Six of the seven judges on the appeals court likened the research to some of the darkest episodes in the history of human experimentation, including the Tuskegee study (see chapter 1), in which treatment was withheld from African-American men with syphilis; the Jewish Hospital study, in which elderly patients were injected with live cancer cells; and even the notorious World War II experiments at Buchenwald concentration camp, in which prisoners were deliberately infected with typhus in order to observe the course of the disease.

"These programs were somewhat alike," Judge Dale R. Cathell wrote, "in the vulnerability of the subjects; uneducated African-American men, debilitated patients in a charity hospital, prisoners of war, inmates of concentration camps and others falling within the custody and control of the agencies conducting or approving the experiments. In the present case, children, especially young children, living in lower economic circumstances, albeit not as vulnerable as the other examples, are nonetheless, vulnerable as well."

Even if KKI had adequately informed the parents of lead risks, Judge Cathell concluded, neither parents nor researchers had the legal right to recruit healthy children for a study that offered them no benefit and put them at real risk of harm. Parents, Judge Cathell wrote, "have no more right to intentionally and unnecessarily place children in potentially hazardous nontherapeutic research surroundings, than do researchers. In such cases, parental consent, no matter how informed, is insufficient."

Dr. Gary Goldstein, the chief executive of the Kennedy Krieger Institute, defended the research in a *New York Times* article written after the Maryland Court of Appeals handed down its decision. He countered that many of the children did benefit from the study. "We were not trying to put children in houses and watch them get lead-poisoned," Dr. Goldstein said. "We did not expect anyone to get lead-poisoned. The point was to show, in a neighborhood where 95 percent of the houses contain lead and 35 percent of the kids have lead poisoning, that with some repairs, you could move into a house like this and stay and not get lead-poisoned." He added: "For the majority of kids in the study, lead levels did go down. To compare this to Tuskegee makes no sense."

Medical ethicist Gregory Pence also compared the KKI study to Tuskegee "in that poor, black people were deliberately recruited for a study where harm to them was foreseen. The harm was rationalized as going to occur anyway, like a study in nature, because such people would likely live in such housing if the study did not occur." Defenders of the study argued that housing conditions were better and the risk of harm lower in these partially renovated homes than in other homes that the children would likely live in were these homes not available. A "Summary of Kennedy-Krieger Lead Paint Study Fact Sheet" from Johns Hopkins emphasized that families recruited for the study "had histories of living in non lead reduced homes," and that all the improvements performed in the study "significantly reduced lead dust levels."

Apart from the issue of actual harm, it must have been foreseen by researchers that at least some minimal lead exposure for children in less abated homes was possible, relative to the control groups living in lead-free homes. The point of the research

LEAD-BASED PAINT IN HOMES

Many industrialized nations banned the use of lead paint inside homes and businesses at the dawn of the 20th century, but it was 1978 before interior lead-paint use was outlawed in the United States. Many U.S. homes built before 1978 contain at least some lead, and if a home tests lead-positive, it is important that the paint be removed or contained—"encapsulated" safely. Careful lead repair (called "abatement") is especially important if young children or pregnant women live in the home.

Prior to its controversial 1993 study, the Kennedy Krieger Institute had shown that special techniques for lead removal are essential to children's health, since the common practices

(continues)

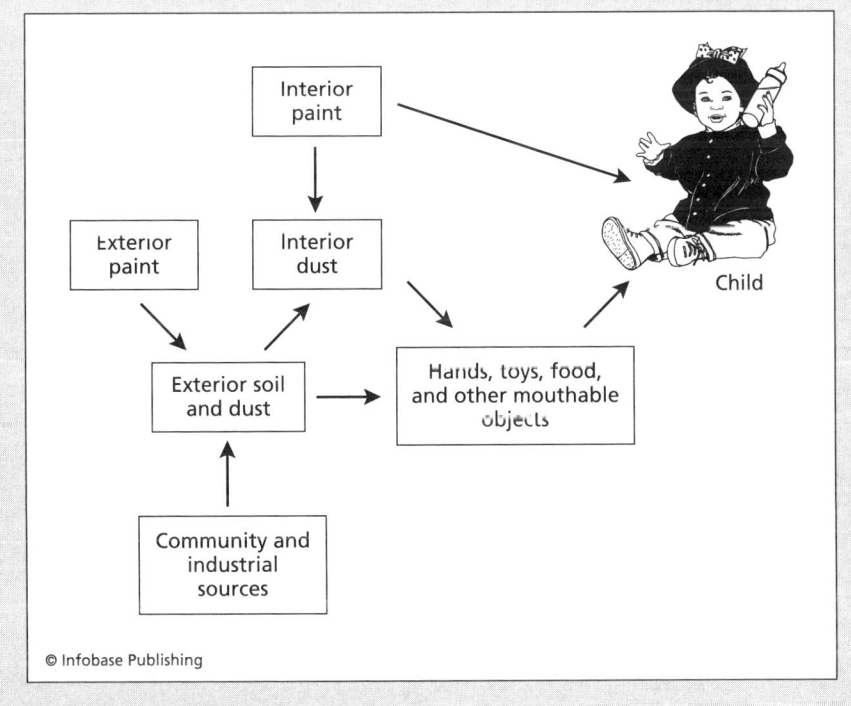

Routes of childhood lead exposure

> *(continued)*
> of burning or scraping only exacerbate the dangers by creating large amounts of hazardous lead dust.
>
> Anyone living in a home built before 1978 (or planning to buy or renovate one) can find testing and safety guidelines at the Environmental Protection Agency's Web site (www.epa.gov). Repairs should not be undertaken without first researching proper procedures for lead abatement and without seeking help from a certified lead-abatement professional. Certified professionals can be located through the National Lead Information Center on the EPA's Web site.

was to perform incomplete lead repairs in order to test the efficacy of less expensive abatement procedures.

Ethicist Alex John London of Carnegie Mellon University proposed that a more instructive comparison than Tuskegee might be the HIV vertical transmission trials conducted in developing nations (see chapter 1). "In both cases the proposed research was designed to be responsive to a health problem that disproportionately affects one population, but not some others. Furthermore, in both cases a major reason for this difference is economic." London acknowledged that the analogy is imperfect, since the countries that hosted the HIV vertical transmission trials were relatively poor, whereas the United States is a wealthy, technologically advanced country, theoretically capable of bringing the safe-housing standards enjoyed by wealthier members of society to all its citizens.

RESEARCH WITH EMERGENCY ROOM PATIENTS: EBB CADE AND THE PLUTONIUM INJECTIONS

During World War II and the cold war, the U.S. government funded a series of medical experiments on humans designed to study the retention of radioactive materials and their effects on the body. Much of the research, and some of the materials themselves, were classified military secrets, and so meaningful consent was rarely, if ever, obtained. Many of the experiments were conducted by doctors in a clinical setting, where patients naturally would have assumed that their health and best interests were of highest priority. Instead, they were being drafted without their knowledge into experiments they did not know existed—experiments that always posed at least a small, but measurable, risk of cancer.

In November 1993, a series of stories in the *Albuquerque Tribune* brought national attention to the human radiation experiments, revealing for the first time the names of patients who had been unwittingly injected with plutonium. Plutonium is a long-lived radioactive material used in nuclear weapons, and the government's rationale for the experiments was the need for data on how it might be retained in the bodies of bomb workers in order to monitor their safety. Soon after the *Tribune* series was published, President Bill Clinton created a presidential advisory committee to review the experiments that were already public, to search for any experiments that might still be classified or undiscovered, and to make recommendations for future research practices.

As the Committee's work proceeded, a major theme emerged from the record: Researchers had used highly vulnerable populations in many of the experiments. Critically ill and comatose patients were injected with plutonium, uranium, and other long-lived isotopes; children and pregnant women were fed radioactive "tracers," or substances used to track metabolic processes; and prison inmates and enlisted servicemen were exposed to external sources of *ionizing radiation*. Some of the patients injected with long-lived isotopes were chosen because they had a *terminal illness* or a life expectancy of less than ten years, so that their chances of developing a radiation-induced cancer were greatly reduced. Others had the potential to live much longer, and were chosen because they had normal metabolisms that would mimic those of healthy bomb workers. Some of the experiments served a dual medical purpose: Attending doctors collected data for the government while hoping that one of the radioactive materials might prove to be a "magic bullet" to target cancer.

The first plutonium injection was amongst the most controversial of all the radiation experiments. Ebb Cade was a 53-year-old African-American man living in the secret nuclear weapons production city of Oak Ridge, Tennessee. Cade, who worked as a cement mixer for a local construction company, had been admitted to the Oak Ridge Army Hospital after an automobile accident on March 24, 1945. He had suffered serious fractures in his arm and leg, but was otherwise "well developed [and] well nourished" and told his doctors that he had always been in good health, according to the experimental record. On April 10, he was injected with 4.7 micrograms of plutonium. This amount probably was much too small to produce any acute toxic effects, but Cade's generally good health meant that his doctors could expect him to live a full life, and therefore they subjected him to a non-negligible risk of radiation-induced cancer.

One Army doctor stationed at Oak Ridge told investigators in 1974 that he had administered the injection to Cade without consent. That doctor's superior, however, told the Committee in 1994 that a different doctor had administered the plutonium. In either event, the likelihood that consent was obtained from Cade for the injection and for invasive follow-up procedures was extremely slim.

Instructions from the Health Division at Los Alamos National Laboratory (another secret weapons complex, in New Mexico) directed the Oak Ridge doctors to obtain blood samples from Cade four hours after the injection, bone tissue after 96 hours, and excretions over the course of several weeks. His broken bones were not set until April 15—three weeks after his accident and five days after the injection—at which time bone samples were removed as instructed by Los Alamos. It is unclear whether doctors postponed setting his bones for the nontherapeutic purposes of the experiment, or whether the injection was timed to occur five days before doctors had intended to set his bones regardless of the experiment. Antibiotics were not as widely available at that time as they are today, and it was common for doctors to delay a surgery if there was any indication of infection. Experimental records indicated that Cade had "marked" tooth decay and gum disease, and that during his hospital stay, 15 of his teeth were pulled and tested for plutonium. The Committee was unable to resolve the troubling question of whether Cade's teeth were pulled mainly for his own medical benefit or to obtain additional tissue samples.

In a 1995 interview with the Committee, the head of the Health Physics Division at Oak Ridge National Lab recalled that one morning, a nurse at the Army hospital opened Cade's door and found that he had disappeared. He had left the hospital suddenly of his own accord. Attempts were made in 1950 to find him in order to perform additional excretion studies and to follow him "for a possible autopsy," but it is unclear whether he was found at that time or whether follow-up studies of any kind were conducted before he died of heart failure on April 13, 1953, in Greensboro, North Carolina.

Instructions for follow-up with other plutonium injection subjects called for those subjects to be told nothing about the purpose of the studies, further supporting the conclusion that consent had never been obtained from Cade.

In its 1995 report, the Committee concluded that the "egregiousness of the disrespectful way in which the subjects of the injection experiments and their families were treated is heightened by the fact that the subjects were hospitalized patients. Their being ill and institutionalized left them vulnerable to exploitation."

PROTECTIONS FOR EMERGENCY ROOM PATIENTS

It is improbable that human experimentation as deeply unethical as the plutonium injections could take place under the current institutional review process. There are, however, serious contemporary concerns about medical research in emergency settings. Patients treated in the field by emergency medical personnel, and patients admitted to emergency rooms, are at their most vulnerable. They are often unconscious or in shock and are frequently alone. It is sometimes hours before loved ones or legal representatives are located who might provide or refuse consent on their behalf.

In 1996, federal regulations were revised to allow for certain types of research on patients in emergency settings without their consent. Under these regulations (21 CFR 50.24), informed consent can be "waived," meaning that patients unable to consent for themselves can be used in research studies—even studies that present more than minimal risk—under certain carefully controlled conditions. There must be "the prospect of direct benefit" to participants, and those participants must be in a life-threatening situation; ongoing attempts must be made to obtain consent from relatives or legally authorized representatives; other available treatments must be "unproven or unsatisfactory"; and it must be impossible to conduct the study without a waiver of informed consent.

Compliance with these federal regulations was at issue in a recent, widely publicized *waived-consent trial* that many critics have called unethical. In early 2006, the research was the focus of a series of articles in the *Wall Street Journal* and an issue of the *American Journal of Bioethics*. In February of that year, the chairman of the U.S. Senate Finance Committee began his own inquiry into the conduct of the trials.

POLYHEME: THE SYNTHETIC BLOOD TRIAL

Imagine that a patient awakes in her local emergency room, only to discover that she has suffered major trauma in a severe car accident; that she went into shock due to blood loss (*hemorrhagic shock*); and that while unconscious, she was enrolled in a "waived-consent" study to test a substance called PolyHeme—a *synthetic blood substitute* derived from donated blood. She received PolyHeme intravenously in the ambulance, where blood—the standard of care for severe blood loss—was not available. She also received the substance in the hospital, where blood was available, even though federal regulations specified that when informed consent has been waived, experimental treatments should only be used when a proven, satisfactory standard of care is unavailable—at the scene of the car accident or in the ambulance, for example.

Most ambulances do not carry blood because of compatibility concerns and difficulties with storage and temperature regulation. Saline solution is used instead to increase blood volume until a patient arrives at the hospital, but saline often cannot increase the delivery of oxygen sufficiently throughout the body. If tissues are starved of oxygen and nutrients for too long, organ failure, and eventually death, will result.

The hope for a synthetic blood substitute like PolyHeme is that, unlike saline, it might be capable of efficient delivery of oxygen to patients. If proven safe and effective, a synthetic blood substitute could ease blood shortages and eliminate the risk of transmitting the viruses that cause hepatitis and AIDS through donated blood. Synthetic blood could prove especially useful to the military, since donated blood is often impossible to carry into combat situations.

In 2003, Northfield Laboratories, the manufacturer of Poly-Heme, launched a nationwide waived-consent clinical trial under a green light from the U.S. Food and Drug Administration (FDA). "This is a groundbreaking clinical trial for a much needed blood substitute and we're very pleased to be involved," said Steven Vaslef, medical director of the Duke Trauma Center and lead investigator for the Duke PolyHeme study site. "Traumatic injury is a leading cause of death in Americans, but no one expects to be injured. It is vitally important to treat trauma patients as early as possible. The treatment process typically begins at the scene of injury, so having the ability to provide treatment with an oxygen-carrying solution may increase the likelihood for survival."

Thirty-two hospitals throughout nineteen states participated in the study, which called for the use of patients under the federal regulations for waived consent in emergency research. The experimental protocol specified that treatment with PolyHeme begin either at the scene of injury or in the ambulance and continue for twelve hours while the patient was in the hospital. It was this latter phase of treatment—the in-hospital phase—that was the focus of intense public criticism, since at the hospital donor blood is the proven standard of care. There were also questions about whether Northfield and participating hospitals failed to warn the public of possible serious risks associated with PolyHeme before beginning the experiment.

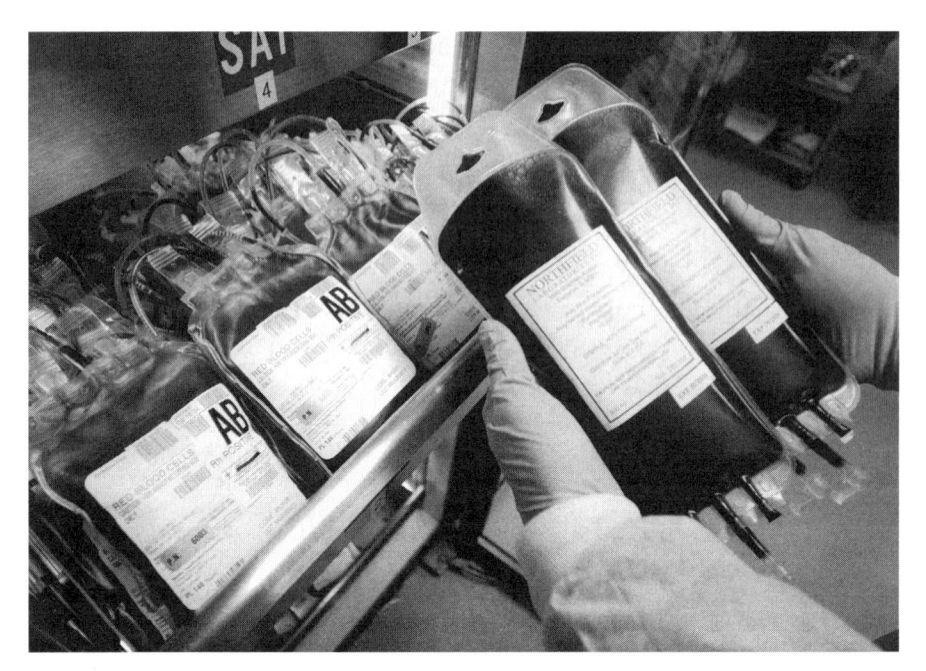

A technician at Northfield Laboratories in Evanston, Illinois, holds IV bags of PolyHeme blood substitute next to expired Red Cross blood. Consent-waived clinical trials showed a slightly higher fatality rate with PolyHeme than with real blood. *(M. Spencer Green/AP Photo)*

On February 22, 2006, the *Wall Street Journal* ran the headline, "Red Flags: Amid Alarm Bells, A Blood Substitute Keeps Pumping." The story revealed that another Northfield trial of PolyHeme had taken place in the late 1990s, the results of which were never made public. The *Journal* obtained internal Northfield documents describing the trial, in which 10 of 81 patients who had received the blood substitute had suffered heart attacks within one week, and two of those patients had died, whereas none of the 71 patients in the control group had suffered heart attacks. The substance was also linked to other serious reactions like abnormal heart rhythm and pneumonia—conditions that occurred in 54 percent of patients receiving PolyHeme, as opposed to 28 percent in the control group.

Northfield had quietly shut that trial down early, concluding in those internal documents that the heart attacks might have been due to doctor error in pushing too much fluid, not a flaw with PolyHeme itself. Individual doctors involved in the trial were told only what had happened to their own patients, and were blocked from knowing the broader results of the study.

Some prominent doctors involved in the earlier trial urged Northfield to publish the results, but to no avail. Dr. Ronald M. Fairman, chief of vascular surgery at the Hospital of the University of Pennsylvania, told the *Journal* that he had repeatedly called Northfield to say, "Let's sit down and write up the data," but Northfield had refused. Dr. T. J. Gan, an anesthesiologist at Duke, said that he also had called Northfield about publication and was told, "Someone's working on it." Dr. Gan said, "Regardless of whatever the problem, you publish it and outline the results." Northfield denied that it "resisted publication." Instead, it said that it "did not allocate resources to publication. In retrospect, reporting the full study results earlier would have been better."

In lieu of consent, the FDA required the hospitals testing PolyHeme to conduct information campaigns at churches, city halls, and other public venues so that their communities would be aware of the trials. In theory, potential patients could opt out of the trials by wearing a wristband stating, "I decline the Northfield PolyHeme study," which they could obtain from the Northfield offices in Evanston, Illinois. How effective these campaigns were at spreading word about the study is in doubt. In July 2006, a reporter with ABC News stood in the middle of downtown Denver—one of the study sites—and stopped people on the street to ask them whether they had ever heard of the PolyHeme study. He was unable to find a single person who was aware of it.

When community members did receive information, it was frequently incomplete. Several hospitals told community meetings that earlier trials had shown PolyHeme to be safe, and their printed materials often failed to mention possible risks, including heart attacks. "Past studies have shown that PolyHeme . . . has not caused organ damage," said one document. Another document from a different locality said, "In clinical trials to date, PolyHeme has demonstrated no clinically relevant adverse effects. Up to now, PolyHeme has not caused any clinically bad problems."

On the heels of the negative publicity, Senator Charles Grassley of Iowa, chairman of the Senate Finance Committee, wrote to Health and Human Services Secretary Michael Leavitt about the study. "I am personally troubled," he said, "that for all intents and purposes, the FDA allowed a clinical trial to proceed which makes the inhabitants" of cities involved in the study "potential guinea pigs, without their consent and, absent consent, without full awareness of the risks and benefits of the blood substitute." He released interagency documents showing that the Office for Human Research Protections (OHRP)—the part of the Department of Health and Human Services charged with protecting human research subjects—had been expressing grave concerns to the FDA for nearly two years about the in-hospital portion of the study.

A senior OHRP official first wrote to the FDA on June 28, 2004, questioning the propriety of giving PolyHeme instead of blood in the hospital when the federal rule clearly states that other available treatments must be "unproven or unsatisfactory." The OHRP official wrote, "There is concern that giving blood to trauma victims is neither unproven or unsatisfactory," and that "due to the gravity of these allegations, the office is concerned that there is an urgency for us to respond." OHRP wrote repeatedly to the FDA over the coming months. In March 2005, the FDA responded that they would require that Northfield "more clearly describe the study" to test communities, but apparently did not address the charge that withholding blood from patients was unethical. In May 2005, OHRP wrote back to the FDA offering to assist in a review of how the informed-consent waiver was being applied, but according to Senator Grassley, it took another seven months for the FDA to respond. By that point, the clinical trial was entering its third and final year.

"Taken at face value," Senator Grassley wrote, "the full series of correspondence between OHRP and FDA, over the course of nearly two years, suggest a breakdown in dialogue within HHS and an apparent disregard by the FDA to zealously fulfill its mission to protect the public health."

The May/June 2006 issue of the *American Journal of Bioethics* also took up the issue of the informed-consent waiver. Kenneth

Kipnis of the University of Hawaii, Nancy M. P. King of the University of North Carolina School of Medicine, and Robert M. Nelson of the University of Pennsylvania School of Medicine, in "An Open Letter to Institutional Review Boards Considering Northfield Laboratories' PolyHeme Trial," said that the research "fails to meet ethical and regulatory standards." They concluded that the field experiment (when the patient was in the ambulance on her way to the hospital) could be conducted in an ethical manner without informed consent, but the in-hospital experiment could not since a proven treatment—donated blood—would be on hand.

Kipnis, King, and Nelson acknowledged that there are well-known immunological problems with blood, that PolyHeme may well have proven to be a better alternative, and that there were good reasons to conduct clinical trials comparing the potential risks and benefits of PolyHeme. As long as it was an unproven alternative, however, and as long as its risks remained unknown, they concluded that it should not have been administered to patients in a hospital setting without their informed consent.

Among other concerns, they noted that "the absence of clotting factors in PolyHeme raises a question whether bleeding secondary to trauma will be adequately controlled in hospitalized patients who receive it instead of blood. PolyHeme could cause deaths in this way (and possibly in other unknown ways)." Considering that some patients on PolyHeme in a hospital setting would inevitably die, if only due to their serious initial injuries, the authors warned that "these men and women will have died while being denied an available treatment (blood transfusions) that is indicated by the standard of practice, following unconsented-to enrollment in a research study."

Despite these concerns, the PolyHeme study—complete with the in-hospital phase—was permitted by the FDA to proceed. It ended in the summer of 2006, having reached Northfield's goal of 720 research subjects—half of whom received PolyHeme instead of blood, even when blood was available. Preliminary results were released in December 2006, and they were disappointing: 46 people had died (13.2 percent) among 349 patients receiving PolyHeme, as compared to 35 who had died (9.6

percent) out of the 363 patients receiving the standard treatment. (Several months later, the company adjusted the results to account for what were termed "discrepancies" in the data, eliminating several patients from the standard-treatment group and identifying an additional death in that group. This brought the death rate in patients receiving standard treatment up slightly, to 9.9 percent.)

Despite the study's disappointing outcome, Northfield's public outlook about its product remained sunny. Northfield's chairman, Steven A. Gould, said that the death-rate findings were not statistically significant. "We're certainly buoyed by the results," he said. Calling the product "noninferior" to saline and blood, the company stressed that its performance had matched the performance of saline in keeping accident victims alive until they could reach the hospital. The experimental protocol was designed in such a way that PolyHeme could be found statistically "noninferior," and still be as much as almost 7 percent worse than treatment with saline and blood in terms of patient deaths.

Whereas Northfield had been permitted to complete its study, an expert panel recommended in December 2006 that the FDA not approve a similar clinical trial for another blood substitute made by a rival company, Biopure—a trial that the U.S. Navy wished to sponsor.

In August 2007, Northfield announced that it intended to apply for FDA approval to market the drug for situations where donor blood was not available. "Given our strong belief," said Gould, "in PolyHeme's ability to address a critical unmet medical need and its potential to save lives, we are committed to making our product available for patients with urgent, life-threatening blood loss when blood is not available."

Amidst the controversy, the FDA convened a public hearing in October 2006 to consider whether the federal informed-consent exception for emergency research should be revised. "There is a certain tension that we have tried to balance between beneficence and respect for persons," said Dr. Sara Goldkind, the FDA's senior medical ethicist, referring to two of the three major ethical principles laid out in the Belmont Report (see chapter 1). "One question we are asking today is, 'Should these tensions be balanced any differently than they are today?'" Many of the experts who spoke at the meeting advocated more thorough attempts to involve and inform communities concerning the risks and benefits of emergency research. Some speakers suggested the formation of a national board to oversee all such research.

Since the 1996 regulations were adopted, about 20 waived-consent emergency research studies have been performed on approximately 2,700 patients. Speaking at the FDA hearing in October, Robert Nelson, coauthor of the *American Journal of Bioethics* open letter to IRBs, said that "failure to conduct a robust and transparent process of community consultation will undermine public trust in the conduct of emergency research."

SUMMARY

Since the days of Tuskegee, Willowbrook, and the plutonium injections, significant advances have been made in the treatment of human subjects. In 1979, the Belmont Report articulated and codified three central ethical principles by which all research with human subjects should be judged: respect for persons, beneficence, and justice (see chapter 1). Local IRBs are charged with scientific and ethical review of experiments, and nearly 10,000 universities, hospitals, and other research institutions have formal agreements in place with the Office for Human Research Protections to comply with federal protections for human subjects.

In practice, however, research with vulnerable populations continues to present special ethical challenges. The outcome of the PolyHeme clinical trial was disappointing, but what if PolyHeme had been found to be an effective blood substitute, with the potential for saving thousands of lives? Would a better outcome have justified placing vulnerable patients at unknown risk without their knowledge? Cases like KKI's lead-paint study and Northfield's blood-substitute trial show that profound tensions between the good of society and the protection of its most vulnerable and disadvantaged members persist in the design and execution of medical research.

3

Research with Animals

Since the mid-1970s, research with animals has produced ample debate in the medical ethics literature about the use of animals as research subjects. People differ greatly on the issue of whether animals should be afforded the same moral status as humans, no moral status at all, or something in between. Some ethicists argue that animals are an extremely vulnerable population—that they are no different in ways that matter from mentally disabled humans. Others argue that animals are missing important human qualities that would afford them equal consideration. This chapter gives an overview of different types of animal testing, looks at a variety of ethical positions on animal research, and reviews two important cases: Harry Harlow's high-profile psychological experiments with monkeys from the late 1950s through the early 1970s, and contemporary research with primates and pigs in the field of cross-species transplantation (xenotransplantation).

An estimated 50 to 100 million animals are used every year worldwide in research conducted inside universities, medical schools, pharmaceutical companies, commercial animal-testing facilities, farms, defense-research establishments, and public-health institutions. Animals are used in basic and applied scientific research, in drug testing, and in science and medical classes. The United States is among countries that still allow animal testing for cosmetic development, though some cosmetic companies in recent years have voluntarily halted animal-testing programs.

In 2006, the U.S. Department of Agriculture (USDA) listed 1,012,713 animal research subjects, according to the following table:

ANIMALS USED IN RESEARCH*	
Species	Number Used as of 2006
Cats	21,637
Dogs	66,314
Guinea Pigs	204,809
Hamsters	167,571
Nonhuman Primates	62,315
Pigs	57,571
Rabbits	239,720
Sheep	13,577
Other Farm Animals	34,632
Other Covered Species	144,567
Total:	1,012,713

* Researchers are not required to report to the USDA on the use of mice, rats, birds, cold-blooded animals (like snakes and frogs), or insects. (Latest available numbers used.)

Animals are subjects in research programs as varied as the species themselves. Some experiments involve no pain; some involve pain without the use of drugs to alleviate it; others involve pain with the use of anesthetic. The following is a brief overview of common research uses for different species. (For more examples, see the "Further Resources" appendix.)

Nonhuman primates (NHP). Because of physical, psychological, and emotional similarities to humans, monkeys and apes are used to study infectious diseases like hepatitis and HIV, neurology, stroke, behavior, cognition, reproduction, genetics, drug abuse, and new vaccines and drugs. Despite their social nature,

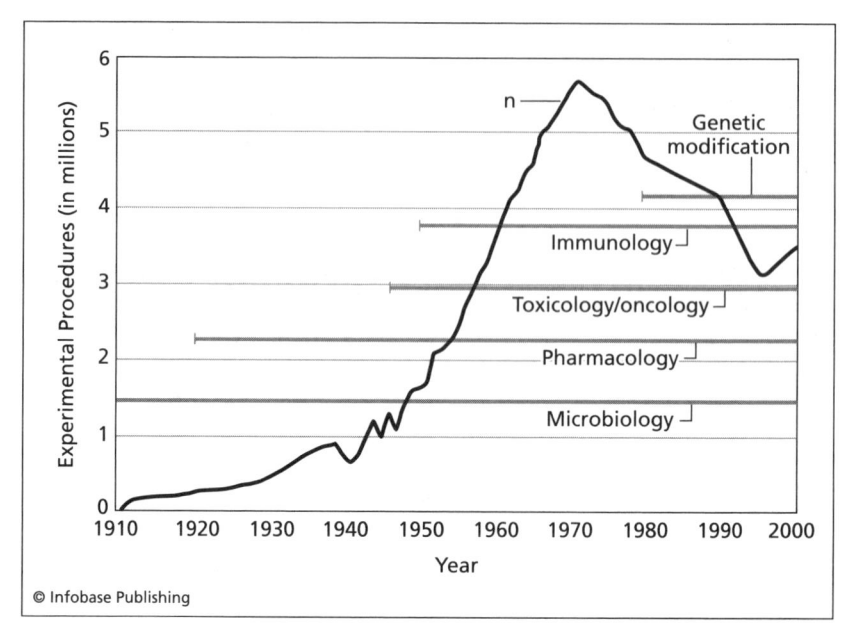

Animal experimentation in the 20th century. The straight lines represent the duration of particular medical research projects. *(Adapted from V. Baumans, "Use of Animals in Experimental Research: An Ethical Dilemma?" Gene Therapy 11 [2004]: 64–66)*

primates are often housed alone due to the nature of the medical conditions being studied. Primate use is on the rise, partly due to xenotransplantion research (the transplantation of an organ from an animal of one species into an animal of a different species). A total of 29,892—or nearly half of all nonhuman primates used in research in 2006—were caused pain or distress. Of these primates, 29,002 received drugs for their pain; 890 did not.

As of 2007, about 1,200 chimpanzees lived in nine U.S. laboratories, down from about 1,800 in 1995. Japan, Gabon, and Liberia are the only other countries currently allowing biomedical testing with chimpanzees. The Humane Society has called for an end to research with chimps, the only "split-listed" animal species—an endangered species also legally available for medical research. Not only are chimps in U.S. labs subjected to invasive experiments, but they can spend a lifetime—up to 60 years—confined to a laboratory.

Research with Animals

Dogs. Beagles are friendly and easy to work with, and are often used in surgery and dental research, as well as in toxicity tests like the LD_{50} (which stands for "lethal dose, 50 percent"). This test determines the dose of a substance or type of radiation required to kill half of test subjects. Due in part to criticisms that the LD_{50} test is scientifically crude and unnecessarily cruel, its use has declined sharply since the early 1970s (replaced largely by the LD_{10} test, which determines the dose required to kill 10 percent of animal subjects).

A total of 30,579—about 46 percent of all dogs used in research—were caused pain or distress. Of these, 29,239 received drugs for their pain; 1,340 did not.

Cats. Cats are commonly used in neurological research. Nearly half of the cats used in research in the year 2006 were involved in experiments causing "pain and/or distress";

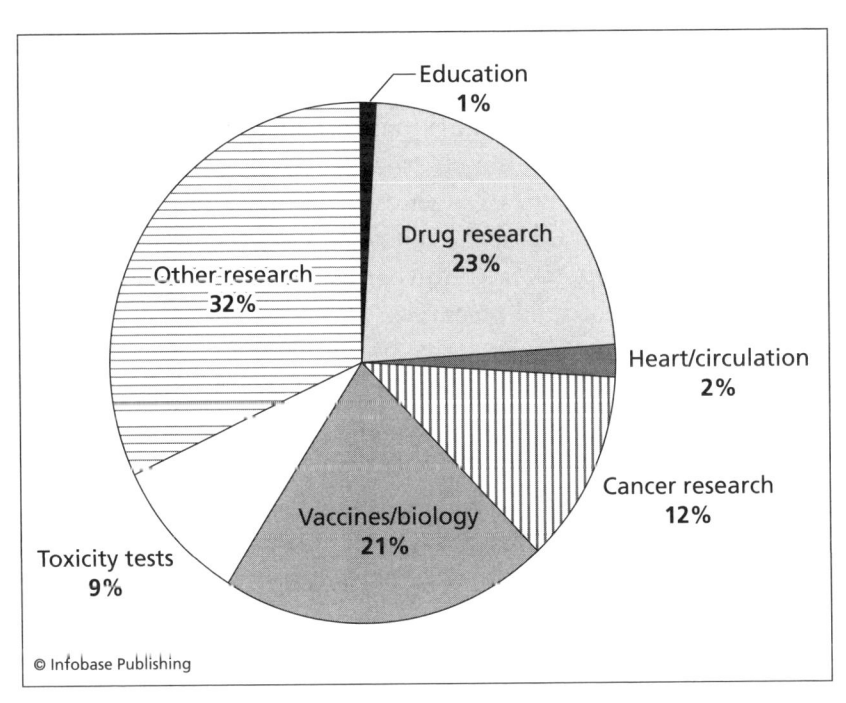

Uses of animal research subjects *(Adapted from V. Baumans, "Use of Animals in Experimental Research: An Ethical Dilemma?"* Gene Therapy 11 *[2004]: 64–66)*

10,018 cats had their pain relieved by drugs, while 412 did not.

Rabbits. Albino rabbits are used in the eye and skin toxicity test called the Draize test (after its inventor, toxicologist John H. Draize). The test involves applying a small amount of a test substance to the animal's eye or skin for four hours and observing the effects for up to two weeks. In the rabbits used, 96,813 were subjected to painful experiments with some relief from drugs, while 6,220 experienced pain with no anesthetic relief.

Mice and rats. These species share 99 percent of their genes with humans and are considered the prime model of inherited human disease. In the United States, the numbers of rats and mice used in research are not reported, but numbers are estimated at 15–20 million.

Insects. Fruit flies are common subjects for genetic studies. The total number of invertebrates used in animal research is unknown, as again, they are not reported to the USDA.

Of the total number of animal research subjects listed for 2006, 577,885 were reportedly used in experiments that did not include more than momentary pain or distress; 382,298 were used in experiments in which pain or distress was relieved by drugs; and 73,640 were used in experiments that caused pain and distress that could not be relieved. These numbers exclude experiments conducted with mice, rats, birds, and insects—the great majority of animal research.

ETHICAL POSITIONS ON ANIMAL RESEARCH

Most people agree that creating unnecessary suffering for any sentient (feeling) creature is something to be avoided. But people often disagree about what counts as suffering, what creatures (other than humans) feel it, and how suffering of humans and other animals should be compared in order to decide how much, and what kind, of suffering is necessary.

Andrew Rowan, executive vice president of the U.S. Humane Society, compares the concept of suffering to another slippery, hard-to-describe concept: "Suffering is usually not defined but the usual implication is that it requires a minimum level of cognitive ability that may not be present in most invertebrates (the octopus being a possible exception). The concept appears to be like obscenity where everybody thinks they can recognize it but nobody can define it for regulatory purposes." Researchers and medical ethicists are thus faced with the sticky problems of (1) how to compare the actual suffering of a lab animal with the potential to relieve human suffering if the experiment yields fruitful scientific results, and (2) how to weigh the suffering of different kinds of nonhuman animals—dogs vs. chimps vs. fruit flies, for example—against their relative usefulness as models of human systems.

Some defining positions have emerged over the past few decades on the acceptability of animal research: Peter Singer's utilitarian arguments against it, and the arguments for and against animal rights.

Utilitarianism

Since the publication of his book *Animal Liberation* in 1975, Australian ethicist Peter Singer has held one of the most controversial and influential positions in the debate over animal research. Singer argues that the vast majority of experiments on animals cannot be justified; that animals are made to suffer in experiments that produce little or no useful knowledge; and that whatever knowledge is garnered could usually be obtained in other ways. He argues that people's willingness to look the other way while animals are harmed is attributable to their *speciesism*, which he defines as a type of bigotry similar to racism and sexism. If experimenters are willing to use animals as subjects, Singer argues, then they should ask themselves whether their research is important enough to use an infant with irreversible brain damage. In his view, if "the experimenter claims that the experiment is important enough to justify inflicting suffering on animals, why is it not important enough to justify inflicting suffering on humans at the same mental level? What difference is there between the two? Only that one is a member of our species and the other is not? But to appeal to that difference is to reveal a bias no more defensible than racism or any other form of arbitrary discrimination."

Singer contends that "adult apes, monkeys, dogs, cats, rats, and other animals are more aware of what is happening to them, more self-directing, and, so far as we can tell, at least as sensitive to pain as a human infant." His approach falls under the broader ethical category of *utilitarianism,* which seeks to maximize good for the greatest number. The first utilitarian to argue against the infliction of suffering on animals was the 18th-century philosopher Jeremy Bentham, who is widely cited by the animal rights movement. "The question," Bentham wrote, "is not Can they *reason*? nor Can they *talk*? but, Can they *suffer*?"

Peter Singer's condemnation of speciesism is echoed in arguments made by animal rights activists. "We may seem like radicals to you," said a member of a break-in team that stole video tapes of an infamous baboon head-injury study conducted at the University of Pennsylvania. "But we are like the Abolitionists, who were regarded as radicals, too. And we hope that 100 years from now, people will look back on the way animals are treated now with the same horror as we do when we look back on the slave trade."

Animal Rights

Most people believe that animals have some rights, contends Andrew Rowan, but most people also believe that they have the right to kill and eat animals. It is a commonly held belief in American culture, grounded in Judeo-Christian teachings, that humans enjoy a divinely-granted supremacy over animals and can do to them as they wish, provided that their behavior is not cruel or reckless. "Thus," writes Rowan, "whatever 'rights' the public believes animals have claim to, they do not include the right to life." The term "right" in philosophical and legal language is typically used to mean a claim that cannot be trumped simply because it would be useful for someone else to do so. For instance, it is widely believed that humans have the right not to be killed for their organs, even if several lives could be saved by transplants resulting from the taking of that one life.

The strongest animal-rights argument, formulated by philosopher Tom Regan in his book *The Case for Animal Rights,* is that nonhuman animals are bearers of moral rights and cannot be used by humans merely as means to the ends of others. Nonhuman animals, Regan argues, are the "subjects-of-a-life"—in other words, they have a life that matters to them just as humans do. If we ascribe value to all human beings regardless of their mental capabilities, then we should ascribe it to nonhuman animals as well. A softer animal-rights argument is that animals have the right not to be caused suffering.

Animals Have No Rights

Some proponents of animal research argue that animals have no rights at all. Philosopher Carl Cohen argues that animals lack the capacity to weigh and balance their own interests against what is right and wrong, and therefore can have no moral claim against others. Cohen goes so far as to say that since it is our duty to prevent risks to human subjects, "the wide and imaginative use of live animal subjects should be encouraged rather than discouraged. This enlargement in the use of animals is our obligation."

Does Middle Ground Exist?

Andrew Rowan argues that despite the "difficult questions, we may find that it is much easier to come to a broadly supported consensus on the ethics of animal research than the overblown rhetoric and [personal] attacks in the media would seem to imply." He presents an alternative to theories like those outlined above—theories that base their arguments on a single morally relevant characteristic such as *sentience* (ability to feel), possession of a life, or the presence of beliefs or desires. In Rowan's view, it is "very likely that no single characteristic is sufficient to describe a complete ethical theory on animal treatment. The world is not that simple. One has to consider a tapestry of characteristics, including the possession of a life, the possession of sentience, the possession of beliefs and desires, the possession of self-awareness, and the like. . . . ." Rowan proposes a multitiered approach in which different animals are afforded certain baseline moral considerations based on characteristics like sentience and self-awareness, and where special obligations are owed to beings with whom we have formed relationships. "Thus," he writes, "one is required to treat one's family with greater consideration than a stranger but the stranger is owed certain basic obligations that cannot be voided."

For animal rights activists, Rowan's approach could prevent many unnecessary and painful experiments on animals with higher emotional and cognitive functioning. For advocates of animal research, it could provide a framework from which to argue for particular experiments that have clear benefits for human health.

HARRY HARLOW AND ATTACHMENT THEORY

Known to his students as "Monkey Man," psychologist Harry Harlow is best remembered for studies of the emotional bond between rhesus monkey mothers and their infants at the Primate Research Center of the University of Wisconsin. Harlow authored the seminal paper, "The Nature of Love," and was elected president of the American Psychological Association in 1958. He remains one of the most controversial figures in 20th-century American science, credited for advancing the now-important *attachment theory* in child development, and—due to the brutal psychological nature of some of his experiments—credited also with generating much of the outrage that has fueled the animal liberation movement in the United States.

Harlow's work helped to debunk widely held beliefs at the time that physical contact between parent and child should be minimized so as not to spoil children—especially male children—and that emotions were not an important part of a child's development. These ideas may seem alien now, but at the time, the behaviorist school of psychology was the dominant voice in child rearing. Feeding, not affectionate parenting, was believed to be the most important source of the mother-child bond. Bill Mason, one of Harlow's graduate students at the time, said, "It's hard to believe now, but when I first started working in Harry Harlow's lab, the prevailing view in psychology was that a baby's relationship to the mother was based entirely on being fed by her."

Harlow's experiments—many of which would be considered unethical by today's animal treatment standards—proved beyond any doubt that for rhesus monkeys, affectionate physical contact

(continues on page 58)

ONCOMOUSE

In the early 1980s, researchers with Harvard Medical School produced a genetically modified mouse that was extraordinarily susceptible to cancer. OncoMouse (from *onco,* Latin for tumor) was created by introducing a gene (*oncogene*) that encourages tumor growth into a fertilized mouse embryo. Harvard requested patents to protect intellectual property rights to OncoMouse from the U.S. Patent Office and from authorities in several other countries. The different outcomes in each case illustrate complex legal and ethical issues surrounding intellectual "ownership" of a life-form and the suffering of *transgenic animals.*

In 1988, the U.S. Patent Office granted patent number 4,736,866 to the President and Fellows of Harvard College for a "transgenic non-human mammal all of whose germ cells and somatic cells contain a recombinant activated oncogene sequence introduced into said mammal, or an ancestor of said mammal, at an embryonic stage." The patent was not limited to transgenic mice: "For example, any species of transgenic animal can be employed. In some circumstances, for instance, it may be desirable to use a species, e.g., a primate such as the rhesus monkey, which is evolutionarily closer to humans than mice."

The legal precedent for granting the patent was a 1980 U.S. Supreme Court ruling that an oil-digesting bacterium was patentable. Chief Justice Warren Burger wrote in the majority opinion that "the fact that micro-organisms are alive is without legal significance" and that patent law covers "anything under the sun that is made by man."

In Europe, Harvard's patent request was eventually granted in 2004, but for different reasons. The key articles of the European Patent Convention excluded patents for inventions "the publication or exploitation of which would be contrary to *ordre public* [public policy] or morality" and patents on "animal varieties or essentially biological processes" for animal production. The European Patent Office (EPO) ultimately decided that Onco-

(continues)

(continued)

Mouse did not count as an "animal variety," and on the issue of the morality exception, developed a utilitarian test to compare the relative harms to the mice and potential medical benefits to humans. The EPO calculated that the potential benefits to cancer patients were greater than the suffering of the mice. Unlike the U.S. patent, however, the European patent was limited to mice.

Notably, a patent request from the Upjohn pharmaceutical company for rights to a different transgenic mouse was denied by the EPO using the same utilitarian standard. This mouse had been genetically modified to lose its hair, and the EPO ruled that the invention's potential utility to research cures for human baldness could not justify causing harm to mice, and therefore denied the patent on the grounds that the invention was contrary to morality.

Two freeze-dried mice, descendants of the first mammals to be patented in the United States. In 1988, Harvard University received a patent for "OncoMouse," a strain of mouse genetically engineered to be susceptible to cancer and especially useful to cancer researchers. *(©SSPL/The Image Works)*

After a series of appeals, the Canadian Supreme Court rejected Harvard's request to patent OncoMouse, concluding that the bodies of mice and other higher life-forms were not patentable because they did not qualify as a "manufacture or composition of matter within the meaning of invention" as intended by the drafters of the Patent Act of 1869. Justices dissenting from the majority opinion argued that the scientific process of changing an animal's genetic material was itself "a composition of matter within the meaning of invention." The court recommended that the Canadian Parliament initiate a public debate to bring some resolution to the contentious subject.

One issue not directly considered in patent proceedings was whether patenting OncoMouse would be good for science. Harvard University signed a memorandum of understanding with DuPont that granted the company licensing rights, and DuPont has since been accused of creating unreasonable barriers to scientific innovation. On June 3, 2002, the *San Francisco Chronicle* reported that leading cancer researchers were accusing the company of "impeding the war on cancer by charging high fees to companies, imposing unusually strict conditions on university scientists and pushing an overly broad interpretation of which lab mice the patents cover."

Peter Shorett, director of programs for the Council for Responsible Genetics, calls patents on genes and organisms a "'toll booth' through which future scientists must pass" and says that the higher the cost of obtaining model organisms like OncoMouse, "the more biomedical innovations will be impeded, as researchers in the early stages of their work may choose to look elsewhere, not willing to pay steep up-front costs or abide by unyielding restrictions." The patent process also hinders a "primary mission" of universities, Shorett argues—the free and open exchange of knowledge. "Secrecy and under-communication become the norm as faculty members withhold data from the scientific community to protect proprietary interests."

(continued from page 54)
between parent and baby was critical to the baby's healthy psychological and physical development. His research served as a foundation for the attachment theory of parenting, for improved treatment of abused and institutionalized children, and for stronger relationships between fathers and their children.

In one experiment, Harlow provided rhesus monkey infants a choice between two artificial "mothers"—a terrycloth-wrapped dummy providing no food, and a wire dummy providing food. The baby monkeys clung to the cloth mother most of the time and spent as little time with the wire mother as possible. If a frightening stimulus was brought into the cage, the baby inevi-

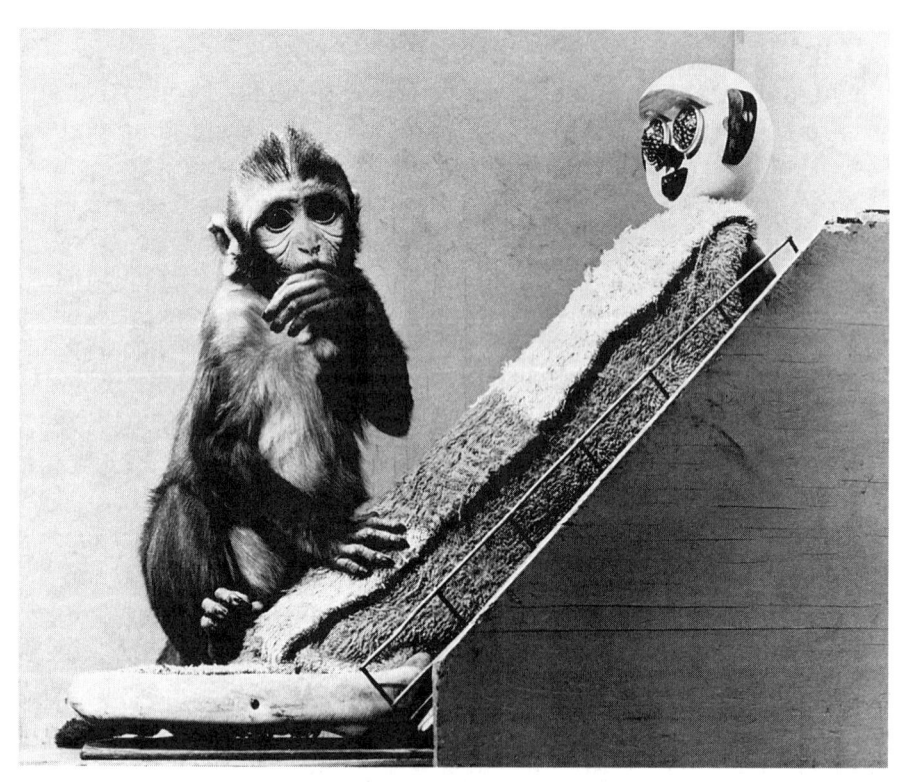

A baby monkey, taken from its mother at birth, sucks its thumb while it clutches an artificial, cloth-covered surrogate mother in one of Harry Harlow's studies of emotional attachment at the University of Wisconsin in the 1950s. *(Hulton-Deutsch Collection/CORBIS)*

tably ran to the cloth mother. When placed in an unfamiliar place with the cloth mother, the baby would cling to her for some time but then eventually begin to explore, returning to her occasionally for comfort. The babies with wire mothers, on the other hand, behaved in a strange new environment as if there were no mother with them at all—crying, cringing, or sucking their thumbs, even running around the strange space searching for the cloth mother.

But Harlow's research did not stop there. In one series of experiments he isolated baby monkeys in steel boxes for a month, six months, or a year. After a month, the "total isolates" were "enormously disturbed" and some starved themselves to death. After a year, the monkeys could barely move, did not explore or play, and were viciously bullied by other monkeys. Some monkeys had to be re-isolated for their own safety. And when Harlow designed a device to forcibly impregnate the females, they proved incapable of parenting their babies, ignoring them or abusing them, sometimes even killing them. "Not even in our most devious dreams could we have designed a surrogate as evil as these real monkey mothers," Harry wrote. "These monkey mothers that had never experienced love of any kind were devoid of love for infants, a lack of feeling unfortunately shared by all too many human counterparts."

Harlow went further, inducing symptoms of *clinical depression* in monkeys by placing them in a *vertical chamber apparatus* he called the "pit of despair." Shaped like an inverted pyramid with walls too smooth to climb, the well was dark inside. Most of the monkeys isolated in this way were already three months old and had formed bonds with other monkeys—in other words, they were socially normal. The point was to break those bonds and see what would happen. The monkeys spent the first few days in the pit trying to scramble up the slippery sides before giving up and assuming "a hunched position in a corner of the bottom of the apparatus. One might presume at this point," said Harlow, "that they find their situation to be hopeless."

In *Animal Liberation,* Peter Singer focused on experiments in which Harlow built "cloth surrogate mothers who could become monsters." One rocked violently, one ejected the baby from its

Speaker **Dog** **Speaker**

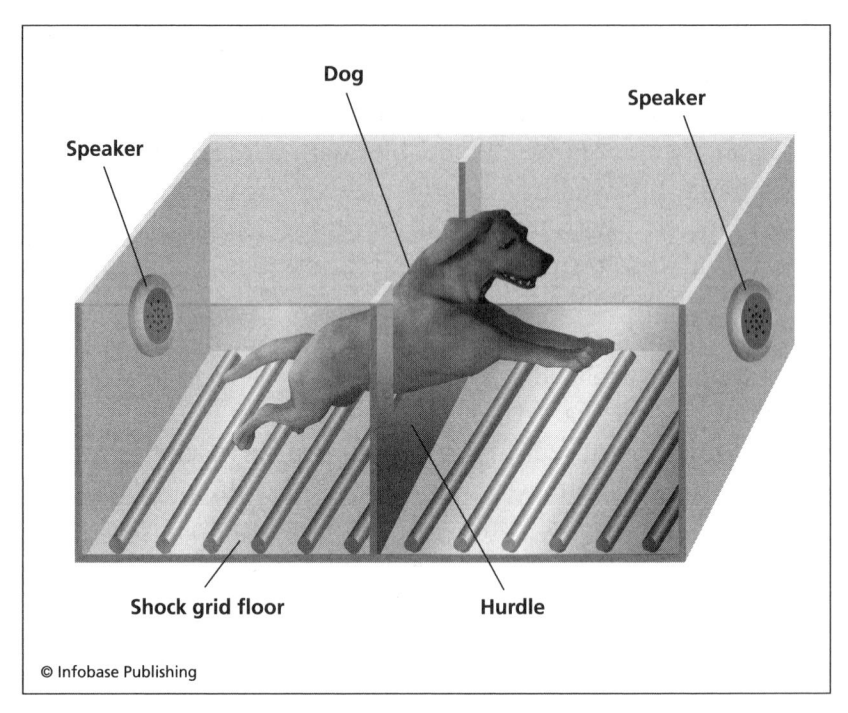

Shock grid floor **Hurdle**

© Infobase Publishing

Sketch of another influential psychological experiment with animals. Martin Seligman's "learned helplessness" theory was based on observations of dogs who failed to learn to avoid electric shock (by jumping over a barrier) after exposure to inescapable shock. The experimental apparatus is known as a shuttle-box.

body, others assaulted the baby with high pressure blasts of air or metal spikes. The babies picked themselves up, waited for the repelling action to cease, and clung again to the cloth "mother." Singer argued that these experiments were torturous and unnecessary—that they taught us nothing that observation of human orphans and other children separated from their parents could not have shown. Harlow's graduate student Bill Mason, now a psychology professor at University of California–Davis, told Deborah Blum, author of the book *The Monkey Wars*, that Harlow "kept this going to the point where it was clear to many people that the work was really violating ordinary sensibilities, that anybody with respect for life or people would find this offensive."

Under a multitiered ethical system like Rowan's, two different kinds of judgments about Harlow's experiments are possible: First, that rhesus monkeys are sentient and self-aware enough that our obligations to protect them from psychological harm should lead us to pursue observational experiments with human children instead; or second, that some of Harlow's experiments could be considered ethical and not others, since some seemed to cause minimal psychological harm to monkeys while improving conditions for human children. Which judgment of Harlow's work is more valid is likely to be at issue for decades to come.

PROTECTIONS FOR ANIMAL SUBJECTS

In 1966, protections for animals were adopted by the federal government in the form of the Animal Welfare Act, a law protecting certain animals in some situations. Not included in the Animal Welfare Act are livestock or other farm animals, horses (except horses used for laboratory research), cold-blooded animals (like snakes and frogs), and insects. In 2002, Senator Jesse Helms introduced an amendment to the Farm Aid Bill that also excluded mice, rats, and birds from the Animal Welfare Act, thus effectively placing the great majority of research animals outside federal protections.

The Animal Welfare Act requires that treatment practices be used in animal research "to ensure that animal pain and distress are minimized, including adequate veterinary care with the appropriate use of anesthetic, analgesic or tranquilizing drugs, or euthanasia"; that alternatives be considered whenever a procedure is "likely to produce pain or distress in an experimental animal"; and that whenever a procedure is anticipated to cause pain, a veterinarian be consulted and painkilling medicines be withheld only when "scientifically necessary" and for "only the necessary period of time." Exceptions to these standards may be allowed if they are "specified by research protocol" and detailed in an annual report filed with the Department of Agriculture.

High-profile abuses of laboratory animals in the 1970s and 1980s led to a Public Health Service requirement that all institutions seeking federal research funds have an Institutional

Animal Care and Use Committee (IACUC) that would ensure that research conformed to the Animal Welfare Act. IACUCs are charged with reducing the number of animals involved, limiting pain and suffering for animals, and using "lower" species rather than "higher" species when the committee deems it scientifically feasible. This approach is based upon the three "Rs" of animal research, first proposed by British scientists William Russell and Rex Burch in 1959: replacement (of animals with "non-sentient material"), reduction (of the number of animals used in research), and refinement (of experiments to minimize inhumane procedures).

Peter Singer has called for local ethics committees that, in addition to carrying out these protective duties, would include animal-welfare representatives who are "entitled to weigh the costs to the animals against the possible benefits of the research," as is the case with animal-care committees in many other countries.

In February 2008, the National Human Genome Research Institute, the National Toxicology Program, and the Environmental Protection Agency announced a joint effort that was hailed by the Humane Society as a major step toward the replacement of animals in chemical testing. The five-year Memorandum of Understanding (MOU) between the three federal entities outlined a cooperative plan to develop robotic molecular and cell-based approaches for determining the toxicity of thousands of chemicals to which humans are exposed.

The MOU states that "a reduction or replacement of animals in regulatory testing is anticipated to occur in parallel with an increased ability to evaluate the large numbers of chemicals that currently lack adequate toxicological evaluation." (The MOU was detailed in the article, "Transforming Environmental Health Protection," published in the February 15, 2008, issue of *Science* magazine.) The Humane Society has gone so far as to say that it believes the new methods should end chemical testing on animals within 10 years.

"This project could eliminate the pain and distress of thousands of animals," said Andrew Rowan, with the qualification that for this vision to be realized within a decade, $200 million per year of funding would be needed, or about four times the current level.

XENOTRANSPLANTATION

It sounds like the plot of a science-fiction movie—"organ farms" that grow genetically engineered pigs in sterile laboratory conditions for the purpose of providing transplant organs to humans. Xenotransplantation (from the Greek "xeno" meaning "stranger") research is the study of cross-species organ transplantation, and it is fueled by the hope that one day a flood of viable organs from animals might alleviate—even eliminate—the severe shortage of transplant organs from human donors. It is a massive research effort currently under way in the United States and other countries.

The first animal-to-human transplants of the modern medical era took place in 1963 at Tulane University in New Orleans, when Dr. Keith Reemtsma transplanted chimpanzee kidneys into 13 patients. Twelve of the 13 recipients died in less than two months, but one survived for nearly nine months. Since that time whole-organ xenotransplants into humans have yielded disappointing results, with the exception of rare cases where temporary transplants from genetically modified animals have prolonged patients' lives until a human organ becomes available. U.S. and world regulatory bodies currently hold that most experimental whole-organ xenotransplants into human recipients are "premature," pending further research with animals.

The majority of whole-organ transplants to date have been performed on monkeys and apes whose organs are removed and replaced with pig organs. Researchers hope that if survival time in nonhuman primates improves, then additional clinical trials in humans might be warranted. The project of xenotransplantation is not without major technical and ethical stumbling blocks, and this chapter will examine three of them: the problem of organ rejection, the risk of creating new infectious diseases, and the question of animal suffering.

Rejection

Perhaps the first attempt to save a human life by xenotransplant was performed in 1902 by Emerich Ullmann, a Viennese physician. Ullmann, who had limited success with animal-to-animal transplants, grafted a pig's kidney to the blood vessels in an arm of a young woman suffering from kidney failure. Artificial kidney dialysis (cleansing of the blood) did not yet exist, and without treatment the young woman was fated to die a prolonged and painful death. As it was, her death came quickly. She did not survive long after the surgery, due presumably to a full-scale attack by her immune system. This type of attack on foreign organs is known as *hyperacute rejection*. Within minutes, or at most a few hours, the unsuppressed human immune system will seek and destroy the organ of a pig or some other non-primate mammal, reducing it to a blackened mass of dead tissue.

Why are pig organs rejected so violently? Most cells extracted from pig tissues do not produce a hyperacute reaction, and some have been used since the 1990s in human trials to research treatments for victims of Parkinson's disease and stroke. Pig heart valves have been used for years to replace malfunctioning human valves and provoke almost no *immune response* in recipients. So why do whole organs trigger such an aggressive attack?

Organs are equipped with blood vessels, and it is the choking of these vessels with clotted blood that ultimately kills an organ. The cells that line the insides of blood vessels, called *endothelial cells,* are the real source of trouble: They are studded with the sugar molecule known as galactose-alpha,1,3-galactose (*GAL*), which the human immune system is already programmed to destroy. Molecules containing GAL also exist on the surfaces of many common viruses and bacteria, and so human immune systems already have antibodies that recognize and attack it.

Biotech companies have been working for years to eliminate this technical hurdle. The recipient's cells are protected by several surface proteins that, in a sense, turn his own immune system away at the door. If the genes responsible for expressing these proteins were inserted into pig *DNA*—creating what

Hyperacute Rejection of Pig Organs

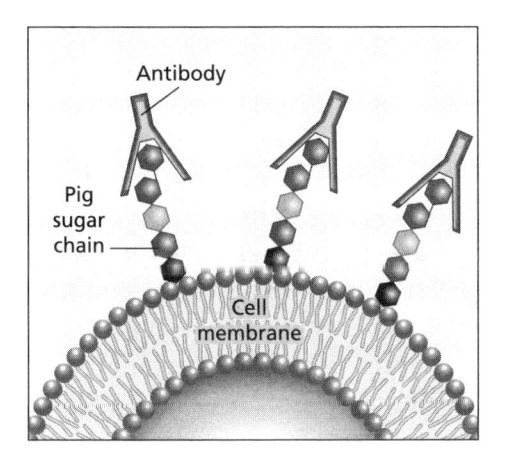

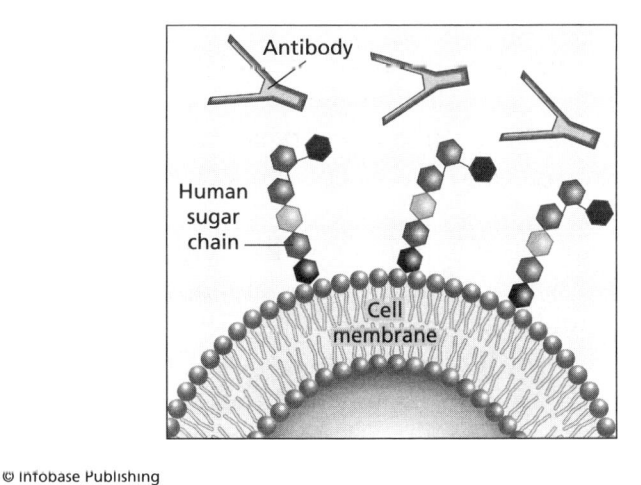

(a) Hyperacute rejection of a pig organ transplanted into a human would likely occur within minutes. The process would begin with antibodies binding to the sugar molecule known as GAL, expressed on the outside of cells that line porcine blood vessels. (b) Organs from pigs genetically engineered not to express the GAL molecule may not elicit such violent reactions.

is called a transgenic animal, whose genetic code is part pig, part human—then blood vessels in the pig organ might also be protected from hyperacute rejection. Transgenic pigs with the protective protein known as DAF on their cells have been created by the xenotransplant company Imutran, and when hearts from these pigs were transplanted into monkeys and baboons, survival times did increase. Still, the hearts were eventually rejected, all within three months.

Another approach is to breed pigs that do not express the GAL sugar molecule on the surface of endothelial cells, and therefore do not invite hyperacute rejection. In January 2002, two biotech companies, PPL Therapeutics and Immerge Bio-Therapeutics, announced production of pigs that lacked one copy of the gene that positions GAL on the endothelial cell surface, and breeding programs removed the sugar molecule entirely. A study conducted at the Transplantation Biology Research Center of Massachusetts General Hospital and published in a 2004 issue of the journal *Xenotransplantation* used these transgenic pig hearts in baboon recipients, with an overall increased survival time and a reported survival time of six months in one animal.

Still a third approach is to modify the recipient's immune system so that it will not reject the pig organ, in effect making the human immune system more "piggish." The idea is to make human recipients into *chimeras*—organisms in which two cell types (in this case, cells from two different species) exist. Chimerism derives its name from the Chimera of Greek legend, a creature that was part lion, part goat, and part serpent. In theory, human chimeras could be created at the cellular level (through *bone marrow* transplants using pig cells) or at the molecular level (through the insertion of pig genes into human DNA). Baboon-pig trials of these techniques have yet to prove successful enough to allow human-pig trials to proceed.

Public Health Risks

Chimpanzee and baboon endothelial cells are not coated in the GAL sugar molecule, and therefore they are not rejected as violently as pig organs. Yet in past human trials with primate organs, the organs were eventually rejected by a slower process known as acute rejection. This process can also occur in human-to-human transplants, but in recent years strong drugs used to suppress the immune system have usually overcome it. Not so with primate organs: No *immunosuppressive drug* yet exists that is powerful enough to prevent cross-species acute rejection. And even if such a drug did exist, other factors have steered research away from the use of organs from apes and monkeys.

In 1999, the FDA banned the use of nonhuman primate organs in xenotransplants due to the risk of cross-species infection. That was the same year that researchers looking into the origins of HIV concluded that the virus probably was passed from chimpanzees to humans through the bushmeat trade, in which chimps and monkeys were hunted and butchered for their meat. In May 2006, HIV-1 M, the most common and virulent form of HIV, was conclusively traced to chimpanzee populations in southern Cameroon, a country in central Africa. (See chapter 7 for a more detailed discussion of HIV and its origins.)

The specter of deadly viruses being passed from apes and monkeys to humans is one reason that in recent years, attention has turned to other animals as potential organ sources, particularly *Sus scrofa domesticus*—the domesticated pig. Pigs share less genetic material with humans than do primates, and many researchers consider them less likely to harbor viruses that might threaten humans. Their organs are the right size to work in the human body, and pigs are domesticated animals—cheap to raise and easy to breed. Killing a pig for its organs is less likely to raise ethical eyebrows than killing a chimpanzee, since many human cultures are accustomed to killing pigs for food.

The use of pig organs is not without public health risks of its own, however. Viruses carried by pigs have proven deadly to people in the past. As recently as 1999, the Nipah virus (named for the Malaysian village where it was discovered) caused *viral encephalitis*—inflammation of the brain—and killed more than 100 pig farmers before it was contained. Scientists are still investigating whether the influenza ("Spanish flu") *pandemic* of 1918—which killed an estimated 20–40 million people—spread to humans through pigs (or to pigs through humans).

Of particular interest to researchers are porcine *endogenous retroviruses,* or *PERV.* A *retrovirus* is made up of a single-strand piece of *RNA* (genetic blueprint) that must be converted to DNA inside a *host cell* in order to replicate. Like the DNA of humans and other mammals, pig DNA contains endogenous retroviruses—ancient retroviruses that long ago worked their way into the species' genetic code. Most of these native retroviruses are not infectious because they are no longer able to function, thanks to the *mutation* of parts of the viral code. But sometimes they can become reactivated, and though they often remain harmless in the original species, they can prove deadly in a new host species. The HIV epidemic is one catastrophic example.

In 1997, virologist Robin Weiss reported in the journal *Nature* that he and his graduate students found three infectious PERV strains in pig DNA, and that at least two of them could infect human cells in culture. The FDA temporarily halted clinical trials using pig tissues, but the moratorium was lifted in January 1998. In 1999, a study conducted by the CDC and biotech company Imutran found no evidence of active infection in 160 patients who had received pig tissues. By 2001, however, new concerns were raised by the work of virologist Daniel Salomon, which showed the first cross-species transfer of PERV during a transplant of pig cells into mice. Salomon and other scientists are careful to note the difference between infection (when PERV infects one cell of the recipient) and *active infection* (if the virus is able to replicate and infect other cells in the recipient's body). Active cross-species infection has not yet been observed with PERV.

The biotech company Immerge found that part of a herd of its miniature swine had organs that would not transfer PERV to human cells, but scientists cannot say with certainty what might happen after an organ has lived inside a human recipient for years. In 2001, Dr. Hugh Auchincloss, Jr., a member of the FDA Subcommittee on Xenotransplantation, said on the public television program *Frontline,* "Could the current pig endogenous retrovirus [PERV], the sequence that defines that virus, change to turn it into a more virulent or different kind of virus? The answer is yes, it could. And we talk about recombination events, where the pig virus joins up with genetic material that's already in the human cells and makes something completely brand-new. These events can occur, we believe, and we don't know what the outcome of them would be."

Critics of the use of pig organs argue that while there are more known viruses from primates, the pig's genetic code has not been studied as well. Alan Berger, executive director of the Animal Protection Institute and former member of the Secretary's Advisory Committee on Xenotransplantation (within the Department of Health and Human Services), told *Newsday* in August 2002 that doing xenotransplantation research with humans is like playing Russian roulette. "There isn't any question that we have a higher degree of known viruses from primates," he said. "But since pigs haven't been studied as well, who knows?" While researchers focus on potential dangers from PERV, other dangers could be missed. There may be viruses no one knows to look for that might become activated in human cells.

Advocates of xenotransplantation counter that the risks of creating a dangerous new human virus is minimal—possibly even less than in human-to-human transplants. Dr. Thomas Starzl, a transplant surgeon with the University of Pittsburgh, told *Frontline* that the risk of creating new infections is "a small one, and with the xenografts, under the conditions that have been stipulated now by government agencies, that risk is very, very remote. And that risk is not in the same league as the risk that you might face" with human-to-human transplants. But critics say that even a remote risk, if it translates into a real event, could represent the deaths of many people. Says Jonathan Allan, a veterinary virologist with the Southwest Foundation for Biomedical Research, "If, in fact, that small risk were to occur, one could still be dealing with tens of thousands of people becoming infected with that virus, maybe thousands dying every year of cancers, or some neurologic disease—something that we hadn't seen before. So out of a small risk in terms of the actual events happening, if it does happen, the potential negative or the harm that could be done could be great."

Transplant surgeon Fritz Bach of Harvard Medical School, medical ethicist Norman Daniels of Tufts University, and others have called for a moratorium on all xenotransplant research pending more public discussion of the risks. Those who want to move ahead with the research argue that without some risk, there can be no progress. But this is a brand-new type of medical risk, Bach and Daniels argue, one that may cause harm not just to the patient but also to the public at large. "We can't quantify the probabilities here," says Daniels, "or the overall magnitude of the payoff and the risk. But there is an enormous uncertainty." And because of that uncertainty, it is wrong, Daniels and colleagues assert, to expose the public to unknown risks of infectious disease without their knowledge and consent. The potential seriousness of the risk, they argue, warrants extensive public discussion.

Bach, who says that apart from these unanswered ethical questions he is one of the strongest supporters of xenotransplantation, argues that now is the ideal time for that discussion. "I'd say that we're in the fortunate position in xenotransplantation where we can't do the organ transplants yet, because we're not ready to. It's a perfect opportunity for [transplant surgeons] to participate with us, to discuss with the public what their feelings are. We cannot proceed, putting the public at potential risk, without doing this. Ethically, we can't. . . . We need to use this space that we have now. It's a wonderful window of opportunity."

Animal Suffering

In May 2000, experimental logs were leaked to the British animals rights group Uncaged Campaigns and to Great Britain's *Daily Express* newspaper. The logs detailed a five-year experiment funded by Imutran, a leading xenotransplant company, and designed to study the survival time of primates with genetically modified pig organs. The logs recorded the condition of the primates morning and evening following the transplants. More than 420 monkeys and almost 50 baboons died, and the average length of survival was 13 days. One-quarter of the primates survived two days or less.

Uncaged Campaigns uploaded the documents to the Internet, along with its own summary of the record, which it called "Diaries of Despair." Uncaged argued that the logs revealed prolonged, unnecessary, and cruel animal suffering and a lack

Animal rights advocates protest outside the British prime minister's residence in London. The demonstrators are opposed to xenotransplantation experiments that remove primates' hearts and replace them with pigs' hearts. *(Sion Touhig/Getty Images)*

of significant scientific progress. The group asked the British government to halt xenotransplants pending an independent judicial inquiry. In September 2000, the *Daily Express* published its own critical account of the experiments, maintaining that "[Imutran] has given a highly selective account of its achievements."

Imutran countered that animals did not suffer and that it had accurately represented its experimental achievements. Citing copyright violation and breach of confidentiality, the company obtained a court order requiring Uncaged to remove the logs from the Internet and preventing them from further reproducing or discussing them. The group is, however, permitted to discuss any information reported in the *Daily Express* article. (The court also forbade the newspaper from further printing or reporting on the documents.)

In a 2001 *Frontline* interview, the director of Uncaged Campaigns, Dan Lyons, was free to talk about one of the experiments detailed in the *Daily Express*. "One of the most unfortunate animals had a piglet heart transplanted into his neck," Lyons said. "It was a particularly disturbing example, I think, because for several days he was holding the heart. It was swollen. It was seeping blood, it was seeping pus as a result of the infections that often occur in the wound site. He suffered from body tremors, vomiting, diarrhea. And the animal just sat there. I think living hell is really the only sort of real way you can get close to describing what it must be like to have been that animal in that situation."

Lyons pointed to early problems with human-to-human transplants that had not been predicted by animal models as evidence that medical advances ultimately depend upon experimentation with humans, not with animals. Xenotransplantation researchers, on the other hand, maintain that experiments with nonhuman primates are absolutely essential to limit risks in future human trials.

SUMMARY

The announcement in February 2008 of a cooperative government effort to develop robotic molecular and cell-based methods to test chemicals for potential toxic effects—and the Humane

Society's assertion that this cooperative program should end chemical testing on animals within ten years—is long-awaited news for animal protection advocates. Chemical testing represents an estimated 10 percent of animal research.

The issue remains, however, whether additional safeguards are needed to ensure that the pain and distress of animals are given appropriate weight in the ethical review process. Xenotransplantation research will continue to be a major battleground for researchers and animal defenders. Regulatory bodies have advised that human trials not go forward until viral risks have proved minimal, and until at least 60 percent of nonhuman primates with pig organs survive three months or more. Unless the project of xenotransplantation is abandoned altogether, it is unlikely that transplant research with monkeys, apes, and pigs will cease in the near future. The search for common ground on the use of animal subjects is likely to fuel debate among researchers, animal defenders, and medical ethicists for years to come.

4

When Life Ends

Since the 1970s, several high-profile medical cases have shone a spotlight on the difficult choices that can arise when technology is used to extend life. Termed "right-to-die" or "right-to-live" cases, depending on perspective, these stories go to the heart of people's scientific, ethical, and spiritual intuitions and can pit doctor against family, judge against doctor, even family against family. This chapter will start with the Karen Ann Quinlan case—the first of many end-of-life stories to play out on the national stage—and will follow the evolution of current biological and clinical definitions of death. The last section will look at the recent fury over Theresa Schiavo's case and what it reveals about the American public's changing relationship to end-of-life interventions.

KAREN QUINLAN

Late one spring night in 1975, Julie and Joseph Quinlan were called urgently to the hospital. Mrs. Quinlan hung up the phone crying. "Karen is very sick," she told her husband.

Karen, their 21-year-old adopted daughter, had been rushed to the hospital by friends who had discovered her unconscious at a party. She was not breathing. She had consumed alcoholic beverages and a sedative, but blood tests later revealed that neither substance was in her system at a level expected to cause unconsciousness.

Whatever mysterious combination of biological factors had triggered her respiratory arrest, she lay now in a coma, breathing but not responsive, receiving oxygen through a mask. After a week like this, she was moved to a more advanced intensive-care facility where testing confirmed her family's worst fears: She had suffered extensive brain damage during the period when she was unable to breathe. Her body had been starved of oxygen for too long, and critical parts of her brain had died. Her family continued to hope for meaningful recovery, but over the course of the next few months, her condition deteriorated. Her body stiffened and her weight fell precipitously to a mere 70 pounds. Soon she was breathing with the aid of a *mechanical ventilator.*

What happened to Karen in the aftermath of that tragic April night would spark a public debate that still rages today. When should life be artificially prolonged? When should care be withdrawn, if ever? What counts as a meaningful life, and who gets to decide? It is no longer extraordinary to see a comatose or brain-dead patient's body maintained on a mechanical respirator or to see a vegetative patient (whose lower brain still functions) receiving life-sustaining nutrients through a tube surgically implanted in his stomach. But cases like Karen's were relatively new in the 1970s. The invention of high-tech life-support devices like the mechanical ventilator allowed the bodies of severely brain-damaged people to be sustained for indefinite periods of time, thus raising ethical and policy questions that remain largely unanswered more than thirty years later.

In July 1975, Karen's parents sought to answer these questions themselves. After consulting with their parish priest, Karen's father asked the hospital to remove her from the *artificial respirator.* At first the attending doctor agreed, but that night he experienced a change of heart. He informed the parents the next day that he could not bring himself to withdraw care. Since Karen was over 21, the Quinlans did not have the legal right to order withdrawal of treatment. They petitioned for guardianship and were turned down in New Jersey Superior Court, but when they appealed, the state Supreme Court decided in their favor. The Court found that parents should be allowed to assert a constitutional "right of privacy" on their child's behalf, and that anyone discontinuing the respirator could not be criminally prosecuted. If such an act caused Karen's death, the Court reasoned, it would not be homicide.

Six weeks later, the hospital had not turned off the respirator, but instead had added another machine to keep her body temperature regulated. After additional pressure from her parents, Karen eventually was "weaned" from the respirator, and, to the surprise of many, continued to breathe on her own. She lived for another nine years in a chronic care facility, receiving high-nutrient tube feedings and antibiotics. On June 11, 1985, 10 years after falling unconscious, Karen died of pneumonia. She was 31 years old.

DEFINING DEATH

Questions like, "When does death occur?" and "How can the moment of death be established with certainty?" have never found easy answers. Some cultures still take extraordinary precautions against premature declarations of death. In the Middle East, for example, there are societies in which a person is not declared dead until three days after his heart stops beating.

Far from clearing up the matter, Western science has tended to complicate the questions. The development of physiology as a field of study in the 17th century allowed scientists to refine their understanding of bodily systems and revealed that the line between life and death was less clear than previously thought. If a rooster's heart could be kept beating by blowing air into the lungs—even after his head was removed—then what was "alive" about the rooster and what was "dead"?

Concerns about premature burial became so widespread in the 19th century—a time when victims of infectious diseases like typhoid fever were buried quickly so as not to risk spreading the disease—that devices were invented to allow for a moving or awakened victim to signal from the grave. Before he died, writer Hans Christian Andersen, best known for his fairy tales, instructed a friend to open his veins before burial to ensure that he would not awake to find himself buried alive.

The invention of powerful new life-extending medical technologies in the mid-20th century complicated the picture

The Premature Burial, by the Belgian romantic painter Antoine Wiertz (1854), depicts a cholera victim awakening and opening his own coffin. Fears of premature burial were widespread in the 19th century. *(Bettmann/CORBIS)*

further. Before the introduction of the mechanical ventilator, patients who were unable to breathe for prolonged periods did not survive. Before the invention of the *PEG tube* and other artificial feeding devices, patients who were unable to swallow wasted away from malnutrition. For loved ones of these patients and their medical practitioners, there were no interventions that would significantly prolong life beyond the natural course of disease or degeneration.

The availability of technologies like mechanical ventilators and *feeding tubes* changed the face of medical treatment and forced people to look at the dying process with unprecedented precision. On the one hand, there was welcome new hope for seriously ill and injured patients who might recover with time. On the other hand, there was a new problem in that these technologies could, in the words of Supreme Court Justice Antonin

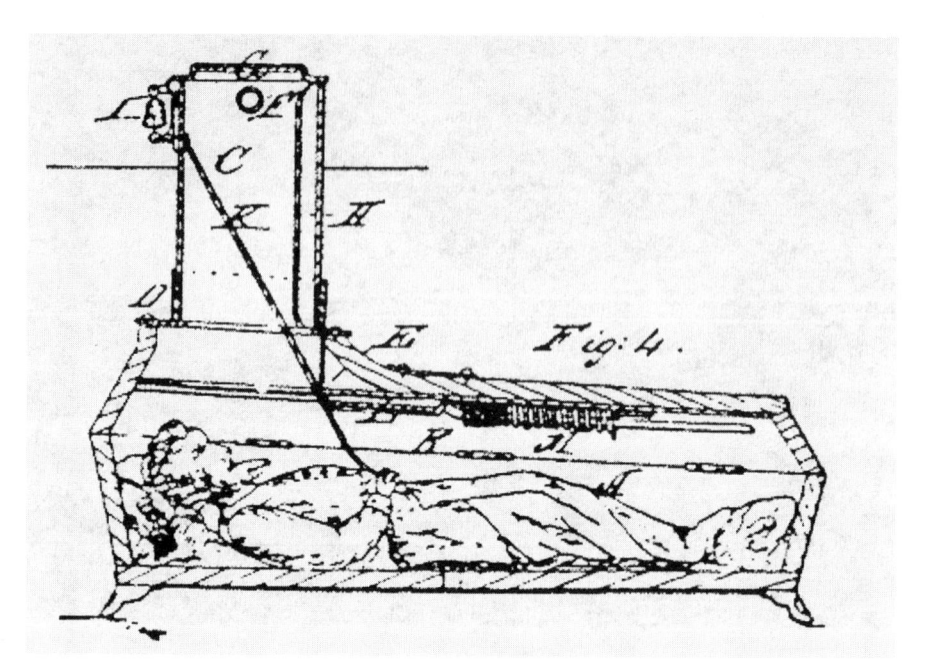

An image of the "safety coffin" from the original patent for the device, which was designed to allow a patient buried prematurely to signal from the grave *(U.S. patent 81,437)*

Scalia in the 1990 Nancy Cruzan verdict, "keep the human body alive for longer than any reasonable person would want to inhabit it."

Medical practitioners, ethicists, lawmakers, families, and patients dissect and debate the meaning of death to ensure that life will be preserved when appropriate and that senseless suffering will be prevented whenever possible. Is death a moment in time or is it a process? If it is a process, at what point should patients, loved ones, or medical practitioners decide that the process has passed the point of no return? Should religious beliefs of the patient, family, or culture-at-large take precedence, or should those beliefs take a backseat to scientific opinion? These questions are posed in every contentious end-of-life case, and since people's answers can differ radically, so can the fates of individual patients.

The most widely noted scientific definition of death, formulated by neurologist James L. Bernat, is the "permanent cessation of the critical functions of the organism as a whole." The "organism" here is understood to be more than a sum of its parts. It is understood to be a unity of complex systems that cannot be further reduced or separated. *Critical functions* are those systems without which the whole organism cannot operate—systems like breathing, circulation, and brain function. Death is defined as the irreversible failure of *all* critical functions. It is sometimes characterized as an event happening at a precise moment in time, and sometimes as a continuous process happening over a period of time. Where exactly the transition between biological life and death occurs is a question science may never answer. But doctors need clinical criteria by which to diagnose a point of no return, a point at which critical functions of the person cannot—or should not—be revived.

Different Kinds of Death?

Terms like whole-brain death, brainstem death, and *heart-lung death* have resulted from the need for observable clinical criteria of death. It is impossible to reliably test a comatose patient's higher brain functioning, for instance, without the use of scanning technologies, so most bedside tests for *brain death* are designed to measure lower brain (brainstem) death.

But are there "real" deaths (e.g., whole-brain deaths) and other "approximations" of death (e.g., heart-lung deaths with residual electrical activity in the brain)? If a person's death is "called" in the emergency room after his heart and lungs have stopped for a critical period of time, or if death is called after he has suffered massive brain trauma in a car accident, the presumption on the part of his doctors is that the failure of either of these critical systems has lead to the same endpoint—that all of his critical functions have failed and that he is no longer alive. Medical ethicist Ronald Munson puts it this way, in his book *Raising the Dead: Organ Transplants, Ethics, and Society*: "Holding that only those who are brain dead are *really* dead is like saying that although you can either drive or fly to Chicago, the real

Chicago is the one you reach by plane. Dead is dead, no matter which set of criteria is employed."

Even if "dead is dead," no two deaths will look exactly alike. The practical need for diagnostic criteria remains, whether death is viewed as an endpoint or as inclusive of all unique paths to that endpoint. Prominent medical ethicist Baruch A. Brody argues in his paper, "How Much of the Brain Must Be Dead?" that since death is a process, and since clinical questions about death get different answers at different points in that process, then, "Rather than seeking a point in the process to serve as the criterion of death and as an answer to these questions, one should choose different points in the process as appropriate answers to the different questions."

Philosophical clarity notwithstanding, clinical criteria are not always cleanly applied in practice. There have been unusual cases where accepted diagnostic tests have been used to establish death, but the patient has spontaneously revived and subsequently recovered. In his 1993 account of his near-death experience, *Raising the Dead: A Doctor's Encounter with His Own Mortality*, surgeon Richard Selzer describes a massive attack of *ventricular tachycardia* (a rapid, life-threatening heartbeat) 23

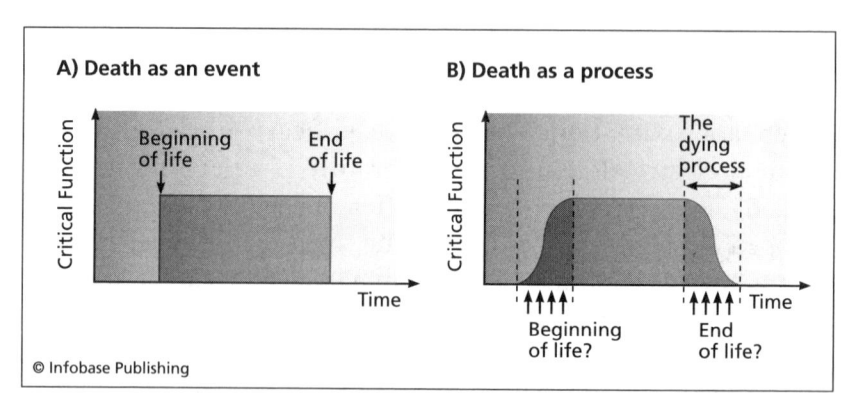

Birth and death depicted as (a) events and (b) processes. Science has not identified testable criteria for determining a precise moment when life begins or ends. *(Adapted from Steven Laureys, "Death, Unconsciousness and the Brain,"* Nature Reviews Neuroscience 6, no. 11 [2005]: 899–909)

days after having been admitted to the hospital for severe pneumonia. Eventually his *electrocardiogram* stayed flat (measured no heartbeat) for over four and a half minutes, and all intervention was stopped. Selzer was declared dead. Ten minutes after "time of death," the attending nurse wrote in his chart that she had observed the characteristic rigidity of death set in. But a moment later he drew a clear breath, the electrocardiogram began recording a heartbeat, and his breathing became regular. When he was later discharged, Selzer says that he told his doctor, "Next time hold a feather to my lips. It's more reliable."

In her book *Twice Dead: Organ Transplants and the Reinvention of Death,* medical anthropologist Margaret Lock describes other cases she calls "narrow escapes." The British newspaper *Guardian Weekly,* for instance, reported the case of a 61-year-old woman who collapsed on New Year's Day, 1996, and was pronounced dead on the scene. Later, in the mortuary, she was observed breathing and was able to recover. In 1989, a 79 year old Canadian man whose brain scans showed "almost" no activity according to his doctors, and who was about to be removed from life support, sat up and stretched out his arms to his grandson. He was described one month later as doing exceedingly well. It is impossible to know with hindsight whether, in these unusual cases, the diagnostic criteria failed to test accurately for death or whether some medical error—human or technological—failed to detect the correct results of those tests. Perhaps Selzer had a weak heartbeat that was not detected, or perhaps his revival defied the standard heart-lung criteria for death.

Transplants, Life Support, and the Search for Clarity

The advent of artificial respiration and high-tech intensive care techniques in the 1950s and 1960s, followed by the discovery in 1972 of *cyclosporine,* a powerful immunosuppressant drug (which lowers a patient's immune response to prevent rejection of a donated organ) made organ transplantation a viable technical option. Then the highly controversial question arose: When is a person "dead enough" for surgeons to harvest her organs?

Removing a donor's organs as soon after her death as possible is a clear and urgent priority for transplant surgeons.

Organs begin to deteriorate quickly after their oxygen supply has been cut off, and the longer the delay in their removal, the more damaged organs become. Surgeons must strike a balance between waiting too long and not long enough, and they have, at times, been accused of rushing organ removal to such an extent that they cause the death of the donor. One U.S. surgeon in the 1960s was charged with homicide, as was one surgeon in Japan. (Typical procedure today is for the transplant team to stay outside the room until the donor is declared dead so as to ensure that there are no conflicts of interest, but a recent case in California has raised questions about whether this policy is adhered to universally; see the section on "Heart-Lung Criteria," page 84.)

Harvard Medical School's Ad Hoc Committee to Examine the Definition of Brain Death was formed in the 1960s to address these practical roadblocks to transplantation. The Committee released its conclusions in an influential 1968 paper, "A Definition of Irreversible Coma," which was published in that year's *Journal of the American Medical Association*. "Obsolete criteria for the definition of death," committee members wrote, "can lead to controversy in obtaining organs for transplantation."

The Committee chairman, Henry K. Beecher, said in 1970 that "Only a very bold man, I think, would attempt to define death." (See chapter 2 for Beecher's role in drawing attention to ethically suspect medical experiments in the 1960s.) Though Beecher's committee did not attempt to define death per se, they did define *irreversible coma* due to brain death as a new criterion for death—where brain death is the irreversible

Influential American anesthesiologist Henry K. Beecher (1904–76), pictured here in 1950 *(Yale Joel/Time & Life Pictures/Getty Images)*

cessation of all functions of the brain. "Function is abolished at cerebral, brain-stem, and often spinal levels," the Committee wrote. "This should be evident in all cases from clinical examination alone."

A second concern of Beecher's was that scarce medical resources, as well as the financial and emotional resources of grieving families, were being spent on patients who would never regain consciousness. The Committee recommended that patients whose brain function had completely and irreversibly stopped, but whose bodies were receiving artificial life support, be pronounced dead and taken off the respirator. The report recommended clinical criteria that should be used to diagnose "whole-brain" death — in other words, the point at which all functioning of the entire brain, both conscious (higher brain) and reflexive (brain stem), had ceased.

Despite widespread belief otherwise, this most prevalent definition of brain death does not include the *persistent vegetative state (PVS),* a state in which the thinking, feeling part of the brain has stopped working but the brain stem still functions. When patients like Karen Ann Quinlan and Theresa Schiavo are called "brain dead," it is a common misperception probably linked to the use of the word "vegetable" to describe brain-dead patients whose bodies are being maintained on ventilators. A surprising 1996 study published in the *Annals of Internal Medicine* showed that nearly one-half of U.S. neurologists and nursing home medical directors surveyed thought that patients surviving in a vegetative state could be declared dead.

Some neurologists and philosophers have argued for a higher-brain definition of death that would link life to human consciousness or "personhood." Under such a standard, permanently vegetative patients could be declared dead and could be buried or otherwise treated as not alive. This highly controversial idea has not been accepted in the mainstream of theory or treatment, in part because there are not well-defined or agreed upon biological markers for consciousness. The permanent vegetative state is still seen by many as something not alive, not dead, but something in between.

The Harvard criteria for whole-brain death served as a model for an influential presidential commission to study the issue, and during the 1970s most states adopted the legal definition of brain death as whole-brain death—"irreversible cessation of all functions of the entire brain."

Heart-Lung Criteria

The Uniform Determination of Death Act (UDDA) is a model law intended to provide consistent, medically sound criteria for establishing death in all cases. It was approved in 1980 by the National Conference of Commissioners on Uniform State Laws in cooperation with the American Medical Association, the American Bar Association, and the President's Commission on Medical Ethics, and has since been adopted by most U.S. states.

According to the act, a person "who has sustained either (1) irreversible cessation of circulatory and respiratory functions, or (2) irreversible cessation of all functions of the entire brain, including the brain stem, is dead." Heart-lung criteria are much more commonly used to declare death than are brain-death criteria, since most people do not die from brain injuries but from illnesses like cancer, *diabetes,* pneumonia, and *heart disease.*

About 98 percent of deaths are declared after the heartbeat or breathing have ceased despite a common belief that patients must be brain-dead before a clear diagnosis is made—though only a small percentage of organ donations come from people declared dead by heart-lung criteria. This percentage appears to be on the rise, according to the United Network for Organ Sharing, as health officials are encouraging more donations after heart-lung death to address the increasingly dire shortage of organs. Heart-lung–death donations reached 5 percent of total donations during the first nine months of 2007—the most for any year this decade.

Donations following heart-lung death have been low in part due to the lack of any legal standard for how long doctors should wait before harvesting organs. The Institute of Medicine has proposed a wait of five minutes—long enough to make spontaneous revival highly unlikely, but also not so long as to severely damage transplant organs—and most donor protocols now call for this wait time. A comprehensive scientific study of spontaneous revival has not yet been performed; such a study would go a long way toward establishing a clear and appropriate standard wait time. In the absence of explicit legal rules for what counts as "irreversible" loss of heart and lung function, it is up to doctors and hospitals to set guidelines for how long to wait.

In February 2008, a transplant surgeon in California was accused of attempting to hasten a patient's cardiac death in order to harvest organs according to heart-lung criteria. If found guilty, the surgeon faces up to eight years in prison. The case has raised fears that much-needed donors and their families will be frightened away, and that other transplant surgeons will face unfair scrutiny. Dr. Goran B. Klintmalm, president of the American Society of Transplant Surgeons, told the *New York Times* in February, "If you think a malpractice suit is scaring surgeons off, wait to see what happens when people see a surgeon being charged criminally and going to jail."

EXTEND LIFE OR HASTEN DEATH?

Dying patients are often fully conscious and competent when faced with end-of-life decisions on their own behalf. A terminally ill patient may wish to stop life support, or she may want to actively end her life early. Should her doctor be allowed to help her die?

Withholding or withdrawing care from a dying patient can be relatively uncontroversial in this country, provided that the futility (uselessness) of treatment is clear to those close to the case. *Physician-assisted suicide (PAS)*, however, is legal only in the states of Oregon and Washington and only under highly restricted circumstances. Oregon's model 1994 law, known as the Death with Dignity Act, made it legal for a physician to write a prescription for a lethal dose of medication for a terminally ill patient. The law prohibits active participation by the doctor in the patient's death and requires the following:

- two doctors agree that the patient has six months or less to live

- if either doctor suspects depression or other mental impairment, they refer the patient for counseling
- the patient make three separate requests for the life-ending drugs (two oral requests 48 hours apart and a written request at least 15 days after the initial spoken request)
- the patient be permitted to terminate the request at any point

Oregon's law is less permissive than the PAS law passed in the Netherlands in 2000, which allows doctors to administer a lethal dose themselves at the patient's request.

Even with the best possible care, dying patients often experience significant pain. A large study of hospital care for terminally ill patients, published in the *Journal of the American Medical Association* in November 1995, reported that "half of the patients who were able to communicate in their last few days spent most of the time in moderate or severe pain." Some patients request a hastened death from their doctors because of pain, or because of any number of other unrelieved physical symptoms or emotional stressors. While PAS is illegal in every state except Oregon and Washington, the U.S. Supreme Court has supported *terminal sedation (TS),* the treatment of pain in terminally ill patients even to the point of causing unconsciousness or hastening death.

TS is the practice of sedating a suffering patient to the point of unconsciousness, usually by means of continuous intravenous administration of a sedative drug. All life-extending measures such as artificial nutrition and hydration, antibiotics, or mechanical ventilation are then withheld, and the patient dies of dehydration or some untreated complication over the course of the next hours, days, or weeks. Because patients are deeply tranquilized, they are thought not to suffer from any pain that might otherwise result from withholding treatment.

While TS and withholding life-extending treatment are legal in the United States, dying patients may not always have access to them, as some medical practitioners may have ethical objections or fears of being prosecuted. Dr. Norman Fost, director of the University of Wisconsin's Program in Medical Ethics, says that despite widespread belief to the contrary, there are no reported cases of doctors in the United States being found liable, civilly or criminally, for withholding or withdrawing life-sustaining treatment from any patient of any age for any reason.

What if the patient is unconscious or mentally incapacitated, as in the Quinlan case? When, if ever, should life-extending care be withdrawn? Many people, at some point in their lives, find themselves in the difficult position of answering these questions on behalf of an ill or injured loved one. People have wildly different intuitions about the answers, as the heated public controversy over Theresa Schiavo's case (discussed in the next section) revealed. Sometimes they find that they have one answer for a stranger and another when forced to decide for a loved one. The prevailing hope, of course, is that the right decision will be clear when it needs to be and that it will be made by a person the patient trusts.

One widely accepted position on life-extending care has been formulated by Roman Catholic ethicists, who draw a distinction between "ordinary" and "extraordinary" care. Ordinary care includes life-extending procedures that are considered standard, offer reasonable hope for improvement, and would not cause extraordinary pain or hardship for the patient. On the other hand, extraordinary care would include any treatment that does not offer reasonable chance of benefit and would cause the patient or loved ones serious suffering.

The distinction is not always straightforward, and different interpretations of ideas like "reasonable chance of benefit" and "extraordinary suffering" are at the heart of most passionate disagreements over whether to administer (or continue to administer) life-extending care. The end-of-life cases discussed in this chapter and the next show that loved ones and doctors—sometimes loved ones and loved ones—can disagree about what constitutes the best care for a patient.

THERESA SCHIAVO

Theresa (Terri) Schiavo, born Theresa Marie Schindler, entered cardiac and respiratory arrest on February 25, 1990, when she was 26 years old. Her husband, Michael, said that he awoke to the sound of her falling around 4:00 A.M. and found that she had collapsed on the floor. By the time paramedics arrived and resuscitated her heart and breathing, her brain had been deprived of oxygen and was severely damaged. She spent the next 10 weeks in a coma and then slipped into the generally unresponsive state where she would spend the next 15 years, sustained by a feeding tube that supplied her body with water and liquid nutrition.

The reasons for her collapse remain a mystery; a low potassium level in her blood, combined with weight loss in excess of 100 pounds since her teenage years, led some doctors to conclude that she was suffering from an *eating disorder* known as *bulimia* and that a resulting potassium imbalance may have caused her heart to stop. The medical examiner who performed her autopsy, however, found no evidence in support of that theory. "The only diagnosis that I know for sure is that her brain went without oxygen," he told the *New York Times* in June 2005. "Why? That is undetermined."

Few end-of-life cases have polarized a country as the Schiavo case did. Two major questions drove the conflicts between her husband and her parents, between conservative Christian politicians and their critics: Could she ever get better? And, since she was no longer able to speak for herself, who should be allowed to speak for her wishes?

Most neurologists who examined her agreed that she was persistently vegetative—that the thinking, feeling part of her brain had died—but her health would be described many ways over the years as the public debate raged around the removal of her life support. Some neurologists near the end of her life said that she might be *minimally conscious,* while other commentators incorrectly described her as brain dead or, at the other end of the cognitive spectrum, as a victim of *locked-in syndrome.*

The following discussion briefly delineates current knowledge of the various levels of awareness:

Brain death is total and permanent loss of all brain function—one of the medical and legal definitions of death (along with heart-lung death). People who have suffered brain death can no longer breathe for themselves, and their hearts will stop beating if they do not receive oxygen from mechanical ventilation.

Coma is a state of profound unconsciousness from which the patient cannot be roused, even by powerful stimulation. Coma may result from a variety of causes including intoxication (by drugs, alcohol, or toxic chemicals), metabolic conditions, diseases of the central nervous system, stroke, head trauma, seizures, and hypoxia (inadequate supply of oxygen).

Persistent vegetative state (PVS) is a state in which the thinking, feeling part of the brain (the cerebral cortex) no longer functions but in which the more primitive part of the brain governing reflexes (the brain stem) still operates. A person in a persistent vegetative state can breathe and sleep but does not communicate or respond to commands in a meaningful way. Vegetative states can sometimes be reversible, though the current diagnostic guidelines allow vegetative states to be declared permanent after six months (in cases of traumatic brain injury) or after three months (in cases like Schiavo's, where brain injury is due to oxygen deprivation).

Rarely, people with oxygen-related brain damage have regained consciousness after being diagnosed as permanently vegetative, but all of those patients recovered within two years. There are astonishing stories of people with traumatic brain injuries who have regained consciousness much later, like Terry Wallis, an Arkansas mechanic, who recovered awareness in 2003 more than 18 years after a serious car accident. Cases like his, many scientists now believe, are more accurately described by the new diagnosis "minimally conscious."

Minimally conscious is a term in use since 2002 to describe some people who formerly would have been described as vegetative but who can track movement with their eyes and seem intermittently responsive. "It took years to get some agreement on the definition," neurologist Nancy Childs told the *New York Times* in April 2005, "and it's only now getting some acceptance, but we've known for years that there was this other group."

Brain scans of minimally conscious patients (see chapter 5 for a discussion of the powerful new *fMRI* scanning technology) show patterns of mental activity similar to healthy patients, with a major difference: The overall rate of energy consumption is lower in the minimally conscious brains than in the normal ones. Neurologists who have studied these brain scans say that they suggest one explanation for the minimally conscious state—that even though much of the neural network seems to be in place (unlike in patients with severe oxygen-related brain injuries, like Schiavo's) there is not always enough energy flow to "light it up" (maintain consciousness).

Locked-in syndrome might be characterized as the opposite of a persistent vegetative state. In this condition, there is severe damage to the brain stem, but the cerebral cortex is unaffected. A person with this syndrome can think and feel emotion but cannot move or communicate except by blinking or eye movement.

There was debate at the end of Theresa Schiavo's life as to whether she was exhibiting signs of a minimally conscious state, but she could not accurately have been called "locked-in." Most neurologists maintained that she was not even minimally conscious. Dr. Ronald Cranford, who examined Shiavo in 2002, said that a brain scan had showed little but scar tissue and spinal fluid, and an *electroencephalogram* (a diagnostic test to record electrical impulses of the brain, or "brain waves") showed no evidence of activity in the thinking parts of her brain. "It's totally flat—nothing," he said, "and this is very unusual. The vast majority of people in a vegetative state show about 5 percent of normal brain activity."

Her autopsy report, released in June 2005, seemed definitive. "This damage was irreversible," the medical examiner said of the injuries to her brain, which had shrunk to half its normal size. "No amount of therapy or treatment would have regenerated the massive loss of neurons." But even after the autopsy, her parents maintained that she had been responsive and that their hope for recovery was reasonable.

This had been their position all along, but Michael and the doctors had disagreed. In the absence of a *living will* (see following sidebar) or any other explicit, written end-of-life instructions,

(continues on page 94)

LIVING WILLS AND MEDICAL PROXIES

A living will or *advance directive* is a legal document expressing a person's decisions about the use of artificial life support in the event that she is unable to speak for herself. This document, or a separate one, may also appoint a family member or friend as a *medical proxy*—someone to make decisions if the patient cannot.

Public interest in living wills and proxies surged in the wake of the bitter controversy over the removal of Theresa Schiavo's feeding tube. A young woman at the time of her collapse, she had no living will or legal proxy, and many people who watched the clashes among family members and public officials saw the need to make their own wishes explicit. Barbara Coombs Lee, president of the Oregon group Compassion in Dying Federation, told the *New York Times* in June 2005, "People are afraid if they don't document their wishes in the most unambiguous way, some politician will try and thwart them."

Aging with Dignity, a nonprofit organization in Florida that works to support people's end-of-life wishes, told the *Times* that in the months following Theresa Schiavo's death, they received 60 times the typical number of requests for their do-it-yourself living-will form. "Mail is coming to us by the truckloads," said Paul Malley, the group's president. Dr. Cecil B. Wilson of the American Medical Association (AMA) reported that 25 times the normal number of people had visited the part of AMA's Web site that addresses end-of-life issues, though he also noted that it would be difficult to predict how many people would follow through on their interest.

Many people who fill out living wills or appoint proxies feel that they do not want artificial measures to be taken that would prolong the dying process, but others are sure that they would want heroic measures. Marilyn Saviola, a polio survivor and advocate for the disabled, told the *Times,* "We're a very disposable society, and I don't want to be considered disposable."

(continues)

(continued)

But research shows that living wills often do not work the way people hope. In a spring 2005 issue of the *Hastings Center Report,* medical ethicist Rebecca Dresser wrote, "As in every end of life case that hits the headlines, people disturbed by the treatment conflict over Theresa Schiavo advised living wills as the remedy. In fact, the importance of having a living will became the main take-home message of this conflict. Unfortunately, living wills are an ineffective remedy."

People change their minds about what they want, and health care proxies often get our wishes wrong. When writing advance directives, people may be motivated by a wish not to burden their loved ones; as circumstances change, however, this wish may give way to other priorities. People often make decisions about their own care with insufficient information, and research shows that their decisions often depend on how the questions are phrased. In one study with elderly patients, for example, 77 percent changed their minds at least once when presented with the same medical scenario but a different description of the possible treatment.

"Living wills call for greater powers to predict circumstances and preferences than most of us can muster," wrote ethicist Carl E. Schneider in the *Hastings Center Report.* "Ms. Schiavo would not only have had to imagine the unimaginable about her physical circumstances, she would also have had to anticipate the ways her social circumstances would affect her medical choices. For instance, had Ms. Schiavo anticipated her parents' inconsolable distress at her death, would she have been willing" to continue on life support?

Across studies of surrogate decision-makers, approximately 70 percent predict their loved one's wishes correctly. Surprisingly, when family members or friends consult a living will, they still only interpret it correctly about 70 percent of the time. A similar result appears with primary care physicians.

Emergency-room doctors, who typically are strangers to the patients they treat, are the lone exception; they are better able to predict their patients' wishes when aided by a living will. They are, however, the least likely of physicians to see the document before treating a patient, since patients with living wills often arrive at the emergency department without them.

Despite periodic waves of interest during high-profile cases—even in spite of a federal law that requires medical providers to give their patients information about advance directives—most people do not have one. In the article "Enough: The Failure of the Living Will," published in a spring 2004 *Hastings Center Report,* Schneider and co-author Angela Fagerlin note that only approximately 18 percent of Americans have signed a living will, and the rates remain surprisingly low for chronically and terminally ill patients. In a study of patients receiving kidney dialysis, for instance, only 35 percent had a living will, though they all thought living wills were a "good idea."

"Living wills were praised and peddled before they were fully developed, much less studied," say Fagerlin and Schneider. Except in cases where a patient's medical crisis is imminent and his preferences clear and strong, the authors advocate the use of medical proxies instead. While proxies are able to accurately predict their loved ones' preferences only about two-thirds of the time, they probably "improve decisions for patients, since surrogates know more at the time of the decision than patients can know in advance."

Another alternative to the living will proposed by Dresser and others is the objective, *best interests standard.* This standard would be applied in cases where patients' past preferences are unclear or absent. In a case like Theresa Schiavo's, in which, according to a judge with the Florida Appeals Court, her "statements to her friends and family

(continues)

(continued)
about the dying process were few" and oral, the hospital would be charged with applying a standard rule about when, if ever, to withdraw care from a vegetative patient, regardless of conflicts among family members. The best response to the Schiavo conflict, writes Dresser, "would be intensified scholarly and public deliberation about the objective standard and its underlying moral judgments." Such a standard, Dresser argues, would include respecting wishes of family members, provided that those wishes are in line with the best interests of the patient.

(continued from page 90)
the most bitterly contested issue became: Who should speak for Theresa Schiavo's wishes?

In the years just after Theresa's collapse, the relationship between her husband and parents seemed strong. "Throughout this period," wrote Jay Wolfson, her court-appointed guardian in 2003, "there was no challenge to either the diagnosis of PVS or to the appointment of Michael as her guardian. For three and a half years, her husband and parents struggled together to maintain her." But in 1993, Michael and the Schindlers began to disagree over the appropriate course of treatment for Theresa, and—according to Michael—her parents demanded to share the malpractice settlement he had won in a lawsuit against her former doctor. By summer of that year, the Schindlers had petitioned to have Michael removed as her legal guardian, and a hostile public feud ensued that would last well beyond her death. "Theresa was by all accounts a very shy, fun loving, and sweet woman who loved her husband and her parents very much," wrote Wolfson. "The family breach and public circus would have been anathema to her."

Eight years after her collapse, Michael petitioned a Florida judge to discontinue Theresa's life support. Her parents fought the petition, but in February 2000, the judge decided in favor of Michael's request. In April 2001, Theresa's feeding tube was removed for the first time. Two days later, another judge ordered it reinserted. Michael went back to the courts and the feeding tube was removed a second time in October 2003. At this time, the Florida legislature passed what became known as "Terri's Law," a law that gave Governor Jeb Bush the power to have the feeding tube reinserted and to appoint a special guardian to review the case. The special guardian, Jay Wolfson, scoured the record and met with Theresa, her family members, and her doctors. He ultimately agreed with the courts that the evidence was "clear and convincing" that she had been accurately diagnosed as persistently vegetative, and that her husband had appropriately represented her wishes and best interests by not further prolonging her life artificially.

In March 2005, "Terri's Law" was ruled unconstitutional by the Florida Supreme Court and her feeding tube was removed for a third time. In a remarkable last-ditch effort to prolong her life, the U.S. Congress passed legislation known as the "Palm Sunday Compromise" that moved the case to the federal courts, and President Bush flew back from his Texas ranch in the middle of the night to sign the bill into law. "In cases like this one," the president said, "where there are serious questions and substantial doubts, our society, our laws and our courts should have a presumption in favor of life." But a federal appeals court in Atlanta refused to allow a new trial or review, and one of the judges admonished President Bush and Congress for acting "in a manner demonstrably at odds with our founding fathers' blueprint for the governance of a free people." The U.S. Supreme Court refused to reconsider the findings of the lower federal court.

Theresa Schiavo died on the morning of March 31, 2005, with her husband at her bedside. She was 41.

SUMMARY

High-profile legal battles over the past 30 years have exposed difficult issues that arise when powerful new medical technologies are used at the end of life. A common theme in the Quinlan case was that medical technology had gone "out of control," and that patients and their families needed to reclaim the "right to die" a peaceful death. By the time Theresa Schiavo's case came to the fore, a shift in the public dialogue had occurred, which culminated in legislative and executive branch intervention that focused on the "right to live."

"Over the last 30 years, the presumption has slowly shifted toward allowing people to die," said medical ethicist R. Alta Charo in the *New York Times* the day after Theresa Schiavo's death. "What we are seeing is the counterinsurgency."

It is unclear how much this shift is reflected in the opinions of the American public as a whole. A Florida poll found that following Schiavo's death, 64 percent of voters disapproved of congressional interference in her case, while 59 percent disapproved of President and Governor Bush's roles in the matter. In various polls leading up to her death, 60 to 70 percent of respondents said they agreed with the decision to remove Schiavo's feeding tube and would advocate the same decision for themselves or a spouse.

Doctors and medical ethicists, however, have observed a shift in the nature of conflicts brought before hospital ethics committees. "About 15 years ago, at least 80 percent of the cases were right-to-die kinds of cases," Dr. Lachlan Forrow, the director of ethics programs at Beth Israel Deaconess Medical Center, in Boston, told the *New York Times* in March 2005. He was referring to cases like Quinlan's, in which families sought to halt what they thought was inappropriate and *futile treatment*. "Today, it's more like at least 80 percent of the cases are the other direction: family members who are pushing for continued or more aggressive life support and doctors and nurses who think that that's wrong."

Apart from the hot-button issue of who should be allowed the final word in end-of-life cases, a central question remains: Can people like Theresa Schiavo know how they might feel years in advance about living in a state of health they have never experienced? Research indicates that they often cannot—that people's preferences for or against heroic measures often change over time—partly because it is impossible to be fully informed about potential medical scenarios, complex life-support technologies, and what it might be like to experience them; and partly because illness and injury can change people's beliefs and their relationships in meaningful and unpredictable ways.

5

Life-Extending Technology

Improvements in medical science and technology have ushered in a new era of hope and of complex choices for patients, loved ones, and health care practitioners. Advances like cardiopulmonary resuscitation (CPR), modern mechanical ventilation, tube feeding, new surgical techniques (such as organ transplants), diagnostic technologies (such as *CT* and *MRI* scans), advanced chemotherapy and antiviral drugs, and modern kidney dialysis machines are all achievements of the past half-century. Chapter 4 looked at clinical knowledge of the dying process and some of the choices that arise from the use of life-extending technology; this chapter will take a more detailed look at four of these powerful new technologies—mechanical ventilation, tube feeding, *neuroimaging* (brain scanning), and transplant technology—how they work to lengthen lives, and how they can impact end-of-life decision-making.

THE MECHANICAL VENTILATOR

This now-familiar device, introduced into critical patient care in the early 1950s, did something revolutionary when it sustained the hearts and circulatory systems of severely brain-damaged people by, in effect, breathing for them. The modern

Life-Extending Technology

mechanical ventilator is a "positive pressure device," which works by expanding a patient's airway with air pressure applied from within, thus causing an inhalation of life-saving oxygen. (The "iron lung"—an older, less effective device—encased a patient's body and pumped the air out around it, causing the chest to expand by the application of "negative pressure.") The positive-pressure mechanical ventilator uses an *endotracheal tube* (a tube inserted into the trachea via the mouth or nose) or *tracheostomy tube* (inserted through an incision in the neck) to introduce gas into the lungs until the "ventilator breath" stops. A natural elastic contraction of the chest then causes the air to flow out, until positive pressure from the ventilator starts the cycle again.

The mechanical ventilator's power to stave off death brought new hope for those who could recover with time and a steady supply of oxygen. The first positive-pressure ventilation case on record is a dramatic example: In 1952, Bjørn Ibsen, a Danish

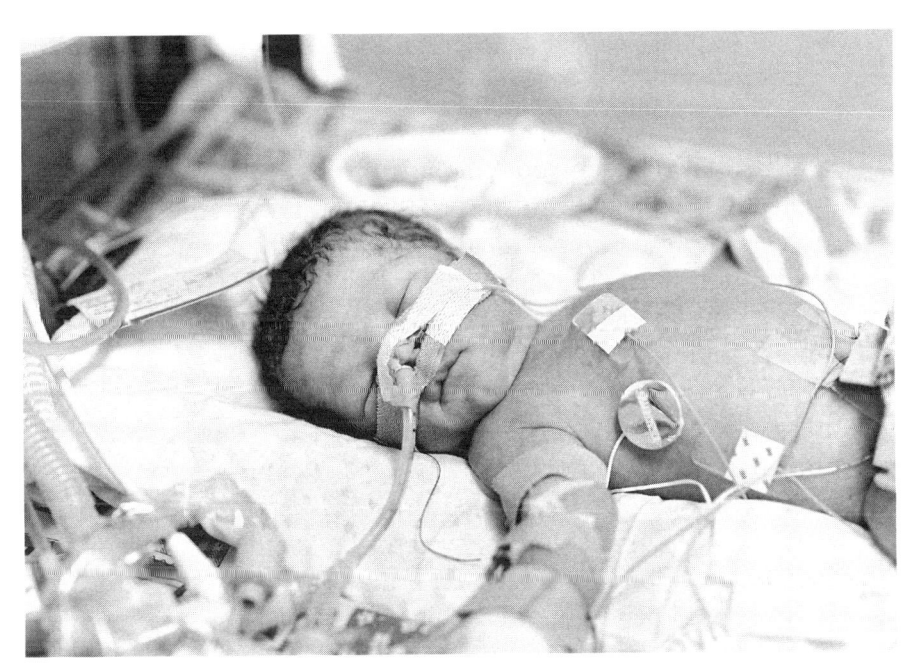

A premature baby receiving oxygen from a modern mechanical ventilator *(Slovegrove/iStockphoto)*

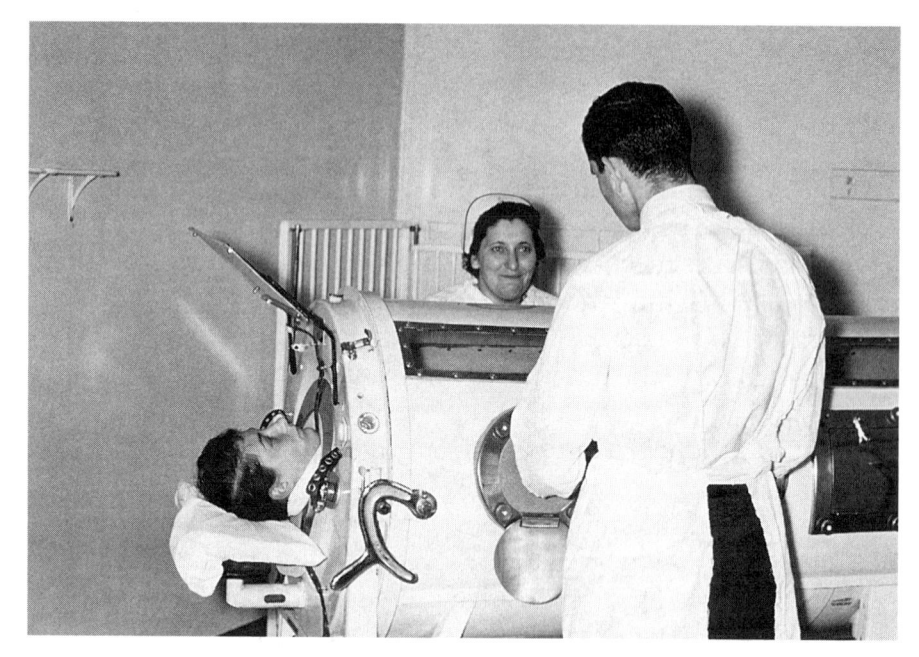

A patient's breathing assisted by an iron lung during the 1950s polio epidemic in Rhode Island *(CDC)*

anesthesiologist, saved a 12-year-old girl whose respiratory system had failed by introducing air into her lungs with a bag he typically used to administer anesthesia. His method was then used to design a mechanical device to do the same work, and the modern mechanical ventilator was born. (See the introduction to this book for a detailed account of Ibsen's life-saving improvisation.)

There was a darker side to the introduction of this extraordinary machine into routine clinical practice. A state of profound unconsciousness known as "irreversible coma" (see chapter 4) first occurred with the ventilator, since before its use patients without working respiratory systems had died from lack of oxygen. Now the bodies of severely brain-damaged and brain-dead people could be maintained indefinitely with a steady supply of oxygen to their living tissues, and loved ones of patients who would not recover were faced with unprecedented emotional and financial burdens.

In the late 1960s, Henry Beecher and the Harvard Ad Hoc Committee to Examine the Definition of Brain Death sought to clarify the situation by recommending that patients in irreversible coma due to brain death be declared dead and removed from the ventilator (see chapter 4). Recent cases like Tirhas Habtegiris's show that the question of when to remove a patient from the ventilator does not always have a straightforward answer, particularly in the minds of loved ones.

When Should Ventilation Be Withdrawn? The Case of Tirhas Habtegiris

In December 2005, Baylor Regional Medical Center in Plano, Texas, removed a woman from the mechanical ventilator against the wishes of her family. Tirhas Habtegiris was a legal immigrant from Eritrea, in northeastern Africa, who was terminally ill with a *metastatic* (spreading) *cancer* that had begun in her abdomen and moved to her lungs. She had been diagnosed late in the summer and had received *radiation therapy* and chemotherapy (see following sidebar) at a different hospital in Dallas, but the cancer was too advanced by that time, and it did not respond to treatment.

At the time of her death, Ms. Habtegiris had been in Baylor's intensive care unit for weeks and had spent most of her stay heavily sedated with morphine to treat severe pain and anxiety. There is controversy over whether she was conscious and coherent at the time that life-extending care was withdrawn. Press reports immediately following her death emphasized the family's claims that she was fully conscious and awaiting her mother's arrival from Africa. Dallas/Fort Worth television station WFAA quoted her cousin as saying, "She wanted to get her mom over here or to get to her mom so she could die in her mom's arms." Contrary to these accounts, the hospital claims that she was never truly conscious, almost from the time of her admission, when doctors began treating her with morphine for severe pain and distress.

The hospital removed her from the ventilator under authority of a controversial Texas law known as the Advance Directives

(continues on page 104)

TAMOXIFEN:
The First Chemopreventive Drug for Cancer

The term *chemotherapy* was coined by Paul Ehrlich a century ago to describe any chemical used to treat a disease. Ehrlich devised a system to test hundreds of heavy metals in the hopes of finding one that would kill the parasite responsible for syphilis, and the 606th compound worked. Known as salvarsan, this substance was the first chemotherapy for any disease, and it cured thousands of patients. The drug's toxic side effects, however, caused illness and even death in some people, and its use was highly controversial. (See the review of the Tuskegee syphilis study in chapter 1 for more on the history of salvarsan.)

Since Ehrlich's time, the development of targeted chemotherapies for cancerous, infectious, and parasitic diseases has saved millions of lives. Some drugs have been discovered by luck, some by systematic testing of hundreds of possibilities, and some by well-planned biological design. Some are tested for one purpose but ultimately prove useful in a different context. This was the case with the pioneering drug tamoxifen, a hormone therapy used for 30 years to treat breast cancer and credited with saving hundreds of thousands of women's lives. It is currently the world's largest selling breast cancer treatment, yet it was not tested initially as a cancer-fighting drug.

The compound ICI46,474, which later became tamoxifen, was discovered by Dr. Arthur L. Walpole, head of the reproduction and fertility control program at the British company ICI Pharmaceuticals Division (now AstraZeneca). The class of drugs to which tamoxifen belongs, known as nonsteroidal antiestrogens, were studied at ICI as possible "morning after" pills, since they had shown contraceptive effects in laboratory animals. Tested in humans, however, they had the opposite effect: They actually induced ovulation in many infertile women, and tamoxifen was then marketed for that purpose.

Life-Extending Technology

Tamoxifen failed as a contraceptive, but Walpole's faith in its potential and his cooperation throughout the 1970s with an American researcher, Dr. V. Craig Jordan, led to its discovery as a breast-cancer treatment. Jordan, who is known as the "father of tamoxifen," began in the early '70s to research the

(continues)

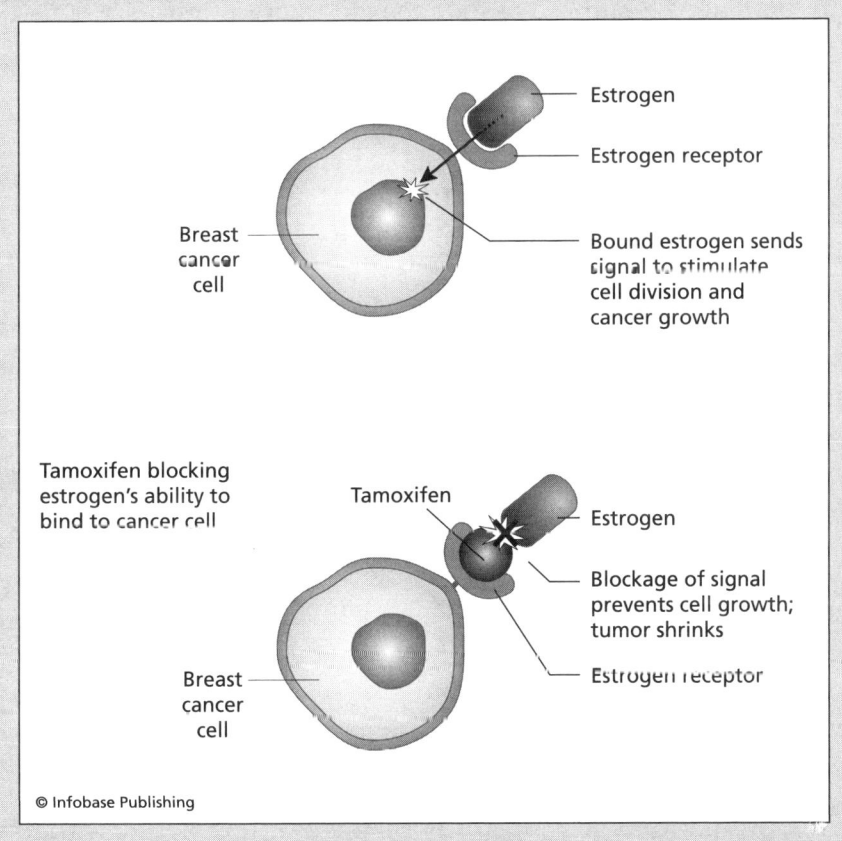

Sketch of how the antiestrogen drug tamoxifen works, depicting (a) the action of the hormone estrogen binding to a receptor site on a cancer cell and helping a tumor to grow, and (b) the action of tamoxifen blocking the receptor site, thus preventing estrogen from binding to the cell

(continued)

idea that tamoxifen might bind to the estrogen receptors (ER) in some breast tumors, thus blocking the absorption of estrogen, which can cause these tumors to grow.

Tamoxifen was approved by the FDA for breast cancer treatment in December 1977 (under the brand name Nolvadex), but it was not a perfect solution. Many tumors—even some *ER-positive* ones—showed resistance to the drug (intrinsic resistance), and some women developed resistance after years of treatment (acquired resistance). Some women experienced undesirable side effects. But as researchers continued to study the way the drug works in the body, there was additional good news: In some women taking it, the incidence of cancer in a second breast was greatly reduced. This finding, along with results showing tamoxifen's preventative effects in mice, led researchers to think that it might be used successfully not only to treat breast cancer, but to prevent it.

A landmark clinical trial conducted between 1992 and 1998 by the National Cancer Institute showed that for women at high risk for breast cancer, taking tamoxifen as a preventive cuts their risk of contracting the disease by almost half. These results led to FDA approval of tamoxifen for breast cancer prevention in high-risk women, making it the first drug to be approved for the prevention of any cancer.

(continued from page 101)
Act (also known as the Texas Futile Care Law), signed by then-Governor George W. Bush. The law allows hospitals to withdraw life-sustaining care that "the attending physician has decided and the review process has affirmed is inappropriate treatment." If a family disagrees, they may request a search for an alternative health-care provider able and willing to provide the life-sustaining treatment. If no such provider is found within 10 days, then treatments other than *palliative care* (treatments to

provide comfort, such as pain medication) may be withdrawn. In Ms. Habtegiris's case, the hospital said that they "contacted 12 facilities including hospitals, long term acute care facilities and nursing homes, all of whom declined to accept the patient."

Ms. Habtegiris was not brain-dead, nor was she comatose or vegetative (as were Karen Ann Quinlan and Theresa Schiavo). She was conscious when not medicated, but according to the hospital, when she arrived on November 15 she was in "severe pain (8 out of 10 on a standard pain scale where 10 is the maximum) and respiratory distress. She was, in essence, actively dying." She was treated with pain medication, and the doses were increased to match her pain, until hospital staff found it impossible to communicate with her. "Every time the doctors or nurses tried reducing the doses of pain medicine and sedatives so they might meaningfully communicate with her, the patient showed evidence of such severe distress that clinical staff had no choice but to restore full sedation."

Press reports highlighted Ms. Habtegiris's lack of medical insurance, implying that it was a determining factor in the hospital's decision to remove treatment. Her brother, Daniel Salvi, told reporters that he believed his sister would never have been removed from life support if she had had health insurance. "If you don't have money in this country," Salvi said, "you're nothing. You're not a human being." Other critics focused on the issue of race, suggesting that while Theresa Schiavo's case made headlines for months, Habtegiris's case did not receive equivalent political or media attention because she was an African immigrant.

The hospital patently denied charges that economics or race had anything to do with their decision to end life support, saying that her medical costs would have been covered by Medicaid had her physicians and the hospital ethics committee determined that prolonging her life was the appropriate course of action. As to charges of racial or ethnic bias, Baylor responded that its professional staff is "ethnically, culturally and religiously diverse," and that when working with patients and families from different cultures, "conflicts may arise." Baylor said that in this case, unlike most other cases involving cultural conflict, its staff was unable to wait for resolution of the dispute due to an urgent duty to relieve the patient's suffering.

On December 12, hospital staff tried one last time to allow the patient to awaken, but her distress was so great that sedation was again increased. With family members and a chaplain in attendance, Ms. Habtegiris was *extubated* and left to breathe on her own. According to the doctor and nurses, her breathing stopped within seconds, and she died "peacefully and rapidly."

TUBE FEEDING

The feeding tube has been at the center of several controversial end-of-life cases. A medical device that bypasses the patient's own swallowing reflex, it can be used in cases of mouth, jaw, throat, or other upper digestive tract injuries or disorders, or in cases of neurologic conditions such as *Alzheimer's disease* or PVS that cause loss of appetite or prevent normal chewing, swallowing, or digestion. When a patient receives nutrition in this manner, it is known as enteral (or tube) feeding.

The first modern feeding tube, termed by one of its inventors the "percutaneous endoscopic gastrostomy" or PEG tube, was introduced in June 1979 when it saved the life of a 10-week-old infant who was unable to swallow. Since that first successful experimental procedure, the use of the PEG tube has skyrocketed in the United States, with an estimated 300,000 people receiving PEGs in 2005—nearly double the number a decade ago. It is a relatively quick and uncomplicated surgical procedure in which a tube is inserted through a small incision in the patient's abdomen and connected to a plastic "bumper," or plug, that has been guided down the patient's throat.

Other means of tube feeding include the *jejunostomy tube,* similar to the PEG tube but inserted into a part of the small intestine rather than the stomach (in cases where the upper digestive tract must be bypassed altogether); and the *nasogastric (NG) tube,* which is inserted into the throat through the nose. The NG tube is suitable only for short-term use, since it tends to be extremely uncomfortable for the patient.

The PEG is the most common artificial feeding device. It was originally intended for emergency care for younger patients, but now is primarily used in older patients with chronic or degenerative diseases. As many as 10 percent of institutionalized older patients are currently tube-fed. Several studies have shown that in general, PEGs in sick, elderly patients tend to cause complications and do not work to extend lives, yet their use in this population continues to grow, perhaps due in part to higher rates of reimbursement by medical insurance and relative ease of use (as compared to alternatives like hand-feeding).

"It spread like wildfire," medical ethicist Stephen Post told the *Wall Street Journal* in December 2005. "Before 1984, nobody with Alzheimer's was on the PEG." But reimbursement for the procedure from Medicare (federal medical assistance for the elderly) nearly doubled between 1988 and 1995. "If you're in a nursing home where a lot of people are PEG fed, that's a poor nursing home," said Thomas Finucane, chief of geriatric medicine at Johns Hopkins.

Whatever has driven the increase, the question of when to withhold or withdraw the procedure is a continuing source of controversy. On one side are doctors and family members who view the PEG as an overused technology that prolongs suffering for frail older or ill patients (such as those suffering from advanced Alzheimer's or existing in a persistent vegetative state). "I've followed very closely where it's gone," one of the PEG's inventors, pediatrician Michael Gauderer, told the *Wall Street Journal*. "It has gone too far."

On the other side of the debate are family members who cannot give up hope, and right-to-life groups who see the failure to use a feeding tube as a violation of religious teachings about the sanctity of life and the duty to use any available means to extend it. Some Catholic ethicists view artificial feeding as a means of "ordinary" care (see chapter 4)—a standard, life-extending procedure that offers reasonable hope for benefit without causing extraordinary pain or hardship. In a 2004 statement, Pope John Paul II affirmed the obligation to feed patients in a vegetative state, calling the administration of food and water "a natural means of preserving life, not a medical act." This view is not uncontroversial in American Catholic health-care circles, where death following the removal of a feeding tube has commonly been viewed as a natural death resulting from a patient's inability to take food normally.

Which of the 50 states a patient calls home may ultimately determine whether or not her feeding tube is removed, as the Cruzan family's battle with the state of Missouri poignantly illustrates.

Cruzan v. Missouri

On January 11, 1983, a 25-year-old woman named Nancy Cruzan was driving on an icy, deserted road in Missouri and lost control of her car, which had no seat belts. The car skidded and flipped, throwing Ms. Cruzan facedown into a watery ditch. When the paramedics found her, she was not breathing and her heart had stopped. Though they were able to resuscitate her, she had suffered irreversible brain damage. Her coma lasted nine months and she awoke to a generally unresponsive state diagnosed as persistently vegetative. Unable to feed herself, she began receiving nutrition through a feeding tube surgically implanted in her stomach.

Over the months ahead her muscles became contorted and rigid and her body curled into a ball. Her husband and parents watched and waited for signs that she might be getting better, but saw none. Her body was being sustained by the feeding tube, but without hope that she would regain mental functioning. Like Karen Ann Quinlan before her, and Theresa Schiavo years later, her brain damage due to oxygen deprivation was too extensive to hope that consciousness would be restored. Her father mourned, "If only the ambulance had arrived five minutes earlier—or five minutes later." Four years after Ms. Cruzan's accident, her parents asked that the feeding tube be removed, but the hospital refused without a court order. The Cruzans went to court.

A county court ruled in favor of the family, finding that "There is a fundamental right expressed in our Constitution as 'the right to liberty,' which permits an individual to refuse or direct the withholding or withdrawal of artificial death-prolonging procedures when the person has no cognitive brain function." The Missouri Supreme Court overturned the lower court's ruling on the grounds that the evidence presented by the Cruzan family that their daughter would not want her life prolonged did not meet Missouri's standard of "clear and convincing" evidence.

The Cruzans appealed to the U.S. Supreme Court, and for the first time in its history, the Court agreed to hear a "right-to-die" case. Though the Court ultimately ruled in favor of the state's prerogative to set its own standards of evidence, the ruling also stated that "for purposes of this case, we assume that the United States Constitution would grant a competent person a constitutionally protected right to refuse lifesaving hydration and nutrition." This was a landmark decision that for the first time upheld an individual's constitutional right to reject unwanted medical treatment, even if the refusal resulted in his death.

Ultimately, the state of Missouri withdrew from the case and the Cruzans went back to county court, this time with additional testimony from friends as to Ms. Cruzan's feelings about life-extending treatment. The case was heard before the same judge who had ruled in their favor two years earlier, and again, the Cruzans won.

Ms. Cruzan's feeding tube was removed on December 14, 1990, nearly eight years after her car crash. Members of the right-to-life group Operation Rescue tried to enter the hospital room to reinsert the feeding tube, but were arrested. Ms. Cruzan died 12 days later on December 26, with her parents, sisters, and grandparents at her side.

NEUROIMAGING

The terms *CT scan* and *MRI* are in common use today. Anyone who has not had one of these procedures has probably heard of them: "Get him up to CT, stat," might be scripted any given night on a fictional hospital television program. Yet these powerful diagnostic technologies have only been available to patients for a couple of decades.

In 1979, Allan McLeod Cormack and Godfrey Newbold Hounsfield shared the Nobel Prize in medicine for their development of computerized axial tomography (CAT or CT scanning), and by the early 1980s, CT scans were widely available for clinical diagnosis. (The CT scanner's existence is partly thanks to the British musical group the Beatles, as their impressive record sales enabled their label, EMI Music Publishing, to fund Hounsfield's research.)

CT technology takes a series of two-dimensional X-ray images and compiles them, through computer algorithms, into three-dimensional images. Uses of the CT scan include diagnosis of *intracranial hemorrhage* (bleeding in the brain), brain tumors, lung changes, *coronary artery disease,* abdominal cancers, and complex fractures.

Advances in computer technology and refinement of mathematical algorithms over the years have vastly improved CT technology. The original CT scanner took 160 images and two and a half hours to reconstruct them, whereas today's scanner can process a 1,000-image series in under 30 seconds.

Magnetic resonance imaging (MRI or MR scanning) was introduced into clinical practice in the early 1980s, and since has been refined and applied at an astonishing pace. Among its developers were Sir Peter Mansfield and Paul Lauterbur, awarded the 2003 Nobel Prize in medicine for their discoveries. MRI technology uses a powerful magnetic field (often up to 30,000 times stronger than the Earth's magnetic field) to align excited water molecules with or against the field. The pattern of absorption and transmission of radio waves by those water molecules is then detected and analyzed by computer.

Functional MRI (fMRI) technology, developed in the 1990s, is a powerful new brain-mapping tool that measures blood-flow changes—and thus neurological activity—by measuring differences in blood-oxygen levels. When brain activity increases, so does the need for oxygen in parts of the brain, and a rush of oxygenated blood causes the MR signal to increase. Functional MRI has come to dominate the brain-mapping field in recent years due to its low invasiveness, relative safety, and wide availability. It is an extraordinarily useful tool in the early diagnosis and treatment of stroke and has played a principal role in the exciting new field of consciousness imaging.

Other types of scanning technologies include: Positron emission tomography (*PET*), which takes biologically active chemicals (such as glucose to detect cancerous tumors, or neurotransmitters like serotonin to study neurological illnesses) and tags them with a radioisotope to measure metabolic (uptake) processes in the body; single photon emission computed tomography (*SPECT*), similar to PET except that it employs gamma-ray emitting radioisotopes and is particularly useful in epilepsy imaging, where an especially rapid "snapshot" of blood flow to the brain is needed; and diffuse optical tomography (*DOT*) which uses light at the near-infrared part of the spectrum (light of a longer wavelength than visible light) to image different levels of oxygen in hemoglobin, the oxygen-carrying protein that gives blood its bright-red color.

Awareness in the Vegetative State

Consider a young woman whose brain was severely injured in a car accident. After a period of time spent in a coma, she opens her eyes and begins to demonstrate sleep-wake cycles. For five months, she consistently fails standard tests to determine whether she is still a thinking, feeling, and aware person. She demonstrates no overt behaviors that could be characterized as willed, voluntary, or responsive. Her doctors diagnose her as vegetative. Then she becomes part of a study to observe brain activity in vegetative patients using a functional magnetic resonance imaging (fMRI) scanner. This newest brain-imaging technology gives neurologists the ability to observe blood flow to working parts of her brain.

The woman is placed inside the fMRI scanner, and something remarkable happens: When she listens to spoken sentences, and then to acoustically matched but meaningless noise sequences, her brain is able to differentiate between the two, lighting up in telltale language-processing patterns in response to the sentences. Adrian M. Owen and the team of British and Belgian researchers studying her case are quick to point out that this in itself is not indisputable evidence of awareness. In a September 2006 issue of *Science* magazine, they emphasize that "many studies of implicit learning and priming, as well as studies of learning during anesthesia and sleep, have demonstrated that aspects of human cognition, including speech perception and semantic processing, can go on in the absence of conscious awareness."

But then a more astonishing result: When asked to respond to mental imagery commands—first, imagining herself playing tennis and then picturing herself walking through the rooms of her home—her brain responds instantly and sustains the mental "work" for a full 30 seconds. On screen, her patterns of mental activity—measured by blood traffic to movement and language centers of her brain—look just like a healthy person's. This extraordinary result leads neurologists to the conclusion that this patient, formerly diagnosed as vegetative, is "beyond any doubt . . . consciously aware of herself and her surroundings."

This is a true story about a young woman in England. She was 23 years old in July 2005 when she was involved in a devastating car accident, and her unexpected mental activity was discovered by scientists using the fMRI scanner half a year later. The implications of the discovery are haunting: How many patients currently diagnosed as vegetative might show similar patterns of mental activity on an fMRI scan?

Currently there are an estimated 25,000 to 35,000 patients in the United States diagnosed as vegetative. The great majority of these cases would be unlikely to show the complex mental activity that the young English patient did, since many patients diagnosed as vegetative have suffered extreme oxygen deprivation and massive loss of brain tissue. The chance of recovery is greater for victims of a traumatic brain injury like the young English patient, whose injuries were severe but localized. Still, this remarkable finding has led some researchers to reevaluate the criteria by which the diagnosis of "vegetative" is made. The authors' assertion that the young woman was "beyond any doubt" consciously aware has proven controversial, but her remarkable case points to a need to develop fMRI and other diagnostic tools to evaluate mental activity even in the absence of external signs of responsiveness.

Owen and colleagues suggest future therapeutic uses of brain-scanning technology in addition to diagnostic ones. The young English patient's demonstrable mental actions, they believe, suggest "a method by which some non-communicative patients,

Life-Extending Technology 113

including those diagnosed as vegetative, minimally conscious, or locked-in, may be able to use their residual cognitive capabilities to communicate their thoughts to those around them by

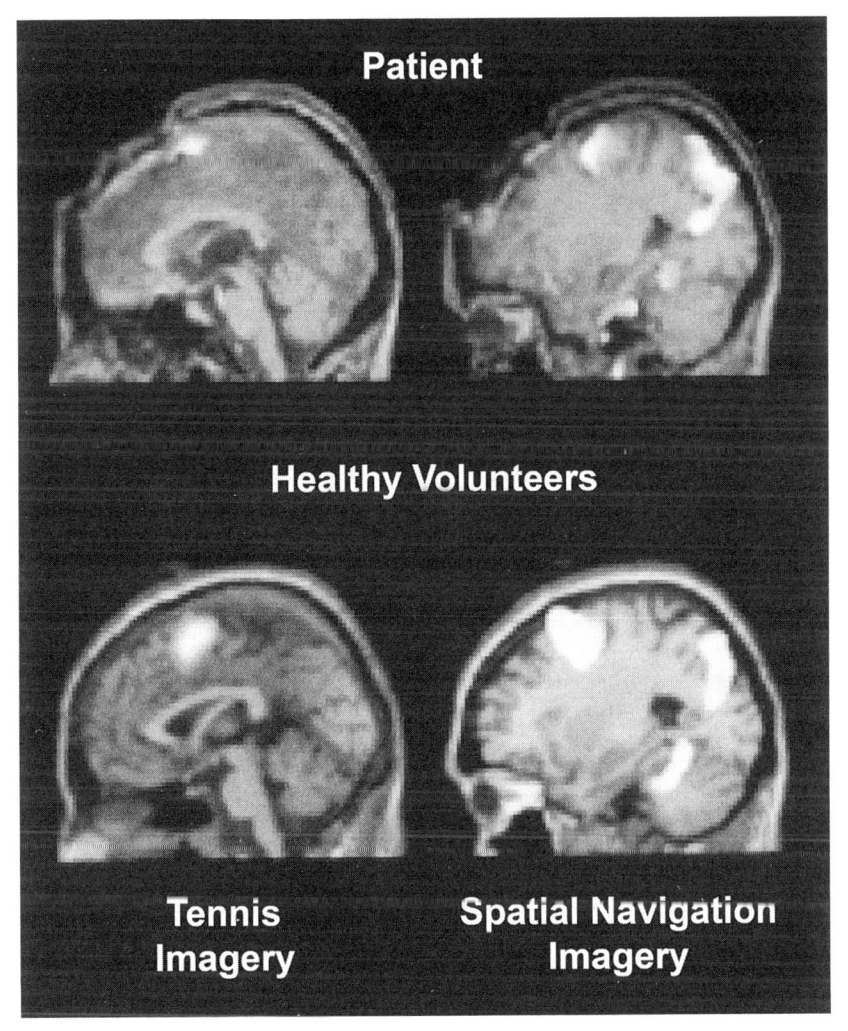

Brain scans of a young British head trauma patient in 2006. The patient was diagnosed as vegetative, but when asked to imagine playing tennis and walking around her house, fMRI images of her brain activity were indistinguishable from those of healthy volunteers. *(Dr. Adrian Owen's Web site, URL: http://www.mrc-cbu.cam. ac.uk/~adrian/Site/Homepage.html. Accessed on July 15, 2008)*

modulating their own neural activity." A patient's thoughts, in effect, could show up on a screen in ways that could be understood by others.

What implications does the young woman's case have for the way we conceive of consciousness and of the line between "alive" and "dead enough" to withdraw care? Chapter 4 outlined diagnostic distinctions among five degrees of consciousness due to brain injury—brain death, coma, persistent vegetative state (PVS), minimally conscious state, and locked-in syndrome—and the young English patient's extraordinary case appears to defy these categories. Nearly 12 months after her accident, her only outwardly observable response to her environment was minimal and had not been reproduced. Researchers described her behavior this way: "In response to a mirror held in front of her, which was then slowly moved to 45 degrees on either side, she turned her eyes very slowly to the right, but not the left, on two trials and fixated for more than five seconds. Thereafter, there was no response to the mirror and no response to any other object, either before or after the trials with the mirror. There was no response to a noxious stimulus except for a transient small dilatation of the left pupil and no response to command."

These results, nearly 12 months after the accident, were ambiguous at best. If her movements were not deemed purposeful or meaningful, she could have been declared "permanently vegetative." If her movements were seen as intermittently responsive, she might have been diagnosed as "minimally conscious"—a diagnosis that would not have reflected the extent of her mental activity.

At the very least, the results of her fMRI called into question the assumed relationship between external signs of responsiveness and mental activity, while answering a question for the young English patient's loved ones—a question asked by family and friends at the bedside of comatose and vegetative patients everywhere—"Can she hear what I am saying?" In this extraordinary case, the answer appears to be "Yes."

ORGAN TRANSPLANTS

The field of organ transplantation has made astonishing strides over the last 40 years, saving and improving the lives of more than 300,000 people in the United States alone. The advent of mechanical ventilators and other high-tech intensive care techniques in the 1950s and 1960s, and the discovery in 1972 of cyclosporine, a powerful immunosuppressive drug (a drug that prevents organ rejection by lowering a patient's immune response), shifted transplantation from the realm of experiment to the realm of life-saving treatment. Newer, more powerful immunosuppressive drugs—like the drug *FK-506,* approved by the FDA in 1994—have improved survival rates even further.

Each year, nearly 30,000 Americans are saved from imminent death by donated organs. Less than 40 years ago, nearly all of those now saved would have died. Most early heart-transplant patients, for example, did not survive beyond six months after receiving their new organ; by 1984, two-thirds of heart recipients survived for five years or more. Now more than 70 percent of heart-transplant patients survive for at least five years.

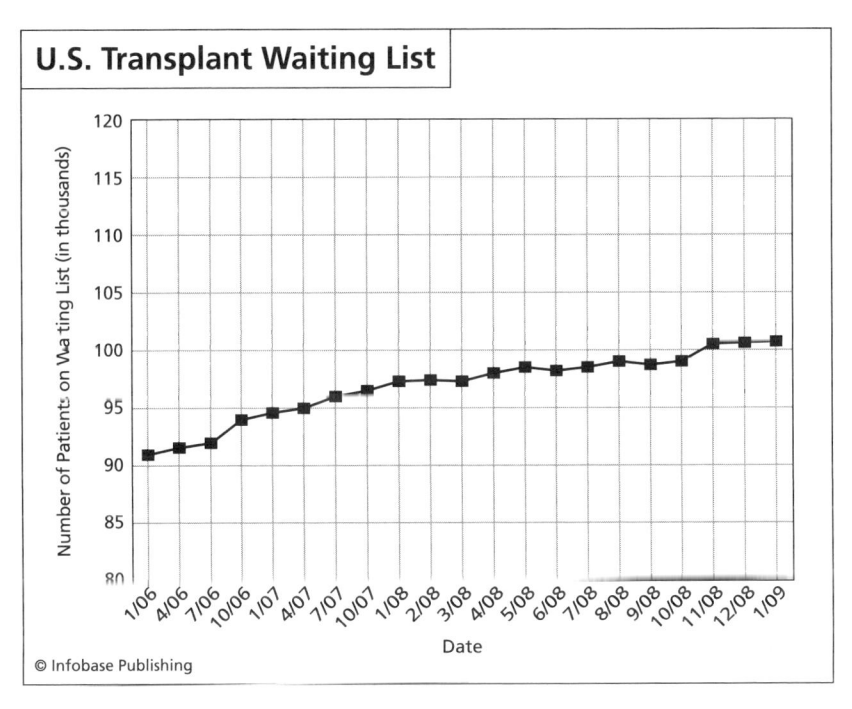

Recent growth of the U.S. transplant waiting list *(Adapted from lifesharers.org)*

About 28,000 people received organ transplants in 2007, yet the waiting list for a transplant has grown at a staggering rate. More than 98,000 people awaited a donor as of March 2008, up from about 50,000 in 1996 and 75,000 in 2001.

Some patients need an organ urgently; others are not as desperately ill, and may wait months or years for a transplant. Most will get the organ they need in time, but many will not: Nearly 6,000 people died in the United States in 2006 awaiting a transplant—16 people per day. Since organs are always in shortage, public policies designed to answer the question, "Who should be next in line?" are inevitably controversial. Any answer to the question necessarily benefits some patients and not others. Should alcoholics, for example, be considered victims of a disease and be allowed to compete equally for liver transplantation? Some ethicists argue yes, some no. In an ideal world, the supply of organs would catch up with the need, and such thorny ethical questions would be moot.

One way to increase the supply of organs is to increase the percentage of potential donors who become actual donors. A 1986 federal law requires hospitals receiving Medicaid or Medicare payments (97 percent of hospitals) to inform families at the time of a loved one's death of their legal right to donate organs, but the requirements have led to only a moderate increase in donations (about 10 percent), due in part to physicians' reluctance to press grieving families for a decision.

Another strategy is to look to living donors to help address the shortage. Many rules governing living donors have been relaxed or eliminated altogether in recent years, allowing donors outside a person's family to make the altruistic decision in favor of a kidney, partial-liver, or partial-lung transplant.

Some physicians and ethicists advocate the passage of clear legal standards dictating when to remove organs from donors declared dead by heart-lung criteria. Such standards might greatly increase the pool of potential donors by allowing surgeons to harvest organs while they are still viable without fear of wrongful-death suits. Currently about 5 percent of all donations come from patients who meet heart-lung criteria for death; the

U.S. Waiting List Deaths v. Wasted Organs, 2008

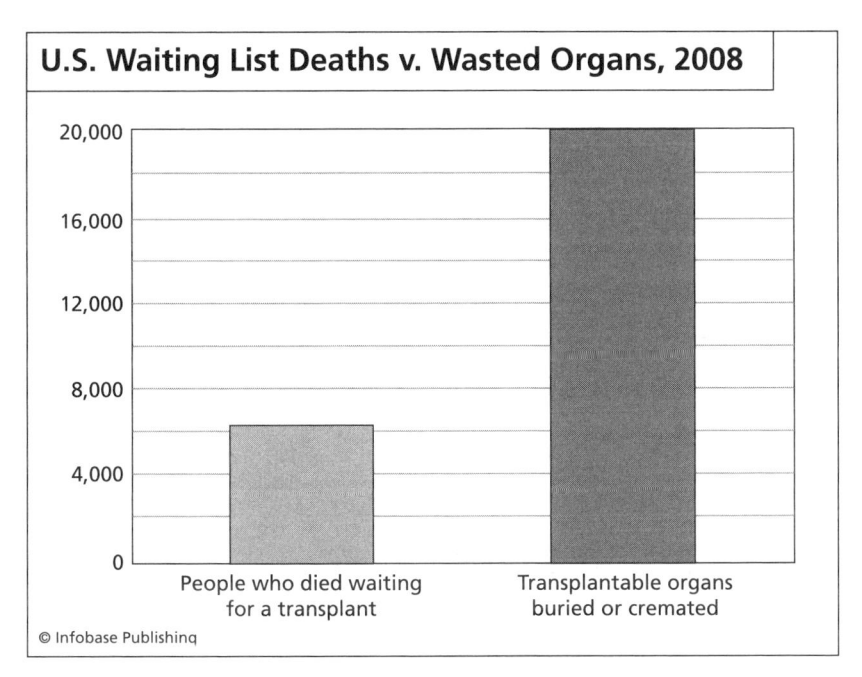

Transplant waiting list deaths v. wasted organs, 2008 *(Adapted from lifesharers.org)*

great majority of donations come from patients who have been declared brain-dead.

A different approach altogether is to try to "grow" transplant organs outside a living human. This is the idea behind xenotransplantation—the transplantation of genetically modified animal organs into humans. (See chapter 3 for an in-depth discussion of xenotransplantation.) Scientists are pursuing other alternatives to animal organs, such as organs grown from *stem cells* obtained from cloned fetuses or umbilical cord blood, or—more recently—from skin cells programmed back to an embryonic state. *Embryonic stem cells* can be used to grow almost any tissue in the body, but the use of fetuses has proven highly controversial in the United States. Embryonic like stem cells produced from skin cells, if they prove effective for this purpose, may be the Holy Grail of organ production. Concerns about the use of fetuses would disappear, and animals would be spared.

Even if a technological solution to the shortage is found, most researchers agree that it will be many years before supply equals demand. In the meantime, real patients and real families face choices that only a few decades ago seemed unimaginable.

Conceiving a Child to Save Another

In 1988, 15-year-old Anissa Ayala was diagnosed with chronic *myelogenous leukemia,* a condition in which the bone marrow produces too many white blood cells. The disease progresses slowly, but if left untreated it is always fatal. Radiation and chemotherapy treatments designed to target cancerous blood and marrow cells typically leave a patient's bone marrow unable to produce enough normal blood cells, and the Ayalas were told that without a bone marrow transplant, Anissa probably would not survive. With a transplant, however, her chances of survival were at least 70 percent.

Anissa Ayala holds her sister and flower girl, Marissa Eve, on her wedding day. Marissa was conceived to provide a lifesaving bone-marrow transplant for her older sister, who suffered from leukemia. Following the transplant, Anissa entered full remission. *(Annie Griffiths Belt/Corbis)*

Neither her parents nor her older brother, Airon, were a "match"—none of them had bone marrow compatible enough with Anissa's to proceed with the treatment. For two years the Ayalas searched nationwide for a compatible donor, but to no avail. They knew that the chances of finding a nonrelative whose bone marrow matched Anissa's were bleak—about one in 20,000.

The Ayalas made the controversial decision to try to conceive another child in the hopes that the child would be a compatible donor for Anissa. Anissa's father, Abe Ayala, admitted that he would not have wanted another child were it not for Anissa's condition. He had had a vasectomy 16 years earlier—and the chances of successful vasectomy reversal go down with time—but against the odds, his surgery worked. Mary Ayala became pregnant at 42, and though there was just a one-in-four chance that the baby's marrow would be compatible with Anissa's, prenatal tests showed that the baby was in fact a match.

The Ayalas were criticized harshly in the press for their decision to have a baby to provide a donor for Anissa. Though bone marrow extraction carries a relatively small risk, some commentators were disturbed that a baby had been conceived to serve someone else's ends without the possibility of its consent. Alexander Morgan Capron, a professor of law and medicine at the University of Southern California, told the *New York Times*, "The ideal reason for having a child is associated with that child's own welfare, to bring a child into being and to nurture it. One of the fundamental precepts of ethics is that each person is an end in himself or herself, and is never to be used solely as a means to another person's ends without the agreement of the person being used."

Anissa Ayala told the *Times* reporter that she was "sort of upset" by criticisms from her parents and that "we're going to love our baby." Mary Ayala said, "Our baby is going to have more love than she probably can put up with." Ethicist James Rachels, in his 1991 *Bioethics* article, "When Philosophers Shoot from the Hip," criticized colleagues who had made "snap judgments" about the Ayalas and a media culture that discouraged any reassurance from ethicists that "alarming events really aren't alarming. That doesn't make good copy."

Rachels wondered if reactions would have been so extreme had Anissa already had a baby sister who happened to be a match. Would the principle that a person should never be used "solely as a means to another person's ends without the agreement of the person being used" prohibit a life-saving transplant because the baby was too young to provide consent? "Should Anissa herself be left to die," Rachels asked, "for the sake of respecting this principle?" He agreed with the Ayalas that what was important was that children, once born, are loved and nurtured by good families, and that in the Ayalas' case, he did not see "how they could have been wrong to weigh their daughter's life more heavily than the philosophers' vague fears."

Marissa Eve Ayala was born on April 6, 1990, and when she was 14 months old, a sample of her bone marrow was extracted and injected into a vein in her sister's chest. The procedure was successful, and Marissa's healthy marrow cells repopulated Anissa's marrow and began to produce new blood cells. After five cancer-free years, Anissa was given a clean bill of health.

Anissa's doctor, Rudolf Brutoco, attended his patient's 1993 wedding. "When you see Marissa walking down the aisle as flower girl," Dr. Brutoco told the *Times,* "you have to realize that either girl wouldn't be here without the other."

SUMMARY

The availability of powerful life-extending technologies like mechanical ventilation and organ transplantation has meant renewed life for many dying patients, and also has required complex decisions from patients, loved ones, and medical practitioners. Sometimes those decisions bring resolution without joy, as in the Cruzan and Habtegiris cases. Sometimes they bring happy results, as in the Ayala case. Sometimes the repercussions of potential life-and-death decisions remain unknown, as exemplified by the young English patient whose fMRI lit up in ways that stumped observers everywhere.

Brain-imaging technology holds the promise of making some of these medical choices less agonizing. By enhancing scientific understanding of various degrees of awareness, neuroimaging may one day clarify when medical interventions are extending meaningful life and when they are inappropriately prolonging death.

6

Life Extension, Aging, and Palliative Care

Americans are getting older. Average life expectancy has soared since the turn of the 20th century, and most Americans see this as a good thing. They plan for a long and prosperous old age and are grateful for advances in nutritional science, new medicines, and life-saving medical procedures that help them to live longer. Many Americans will be able to live well into their 80s or beyond—provided that they are aided by some combination of forces like healthy lifestyle, good access to health care, fortunate genetic predisposition, optimistic outlook, strong social networks, and mothers who were able to stay healthy while pregnant.

The relatively new scientific discipline of *biogerontology* seeks to stretch human lives even longer. Researchers in this field have set out to investigate the biological processes that drive aging and how these processes might be slowed—and life radically extended. The first part of this chapter looks at recent discoveries that might hold the potential to double the human life span and at conflicting ethical positions on whether this is a good thing.

Anti-aging research may or may not drastically alter the human life span, but the past century has already seen American lives lengthened by more than 50 percent on average. The latter part of this chapter examines trends and disparities in life expectancy—nationally and globally—and the growing need for quality, affordable care for an aging population.

THE SCIENCE OF LIFE EXTENSION

Before the 1970s, biogerontology and the idea of *prolongevity* (extension of the human life span) lacked credibility in many scientific circles, and biogerontologists had to work to dissociate themselves from fraudulent projects that claimed false antiaging powers. In 1974, the National Institute on Aging (NIA) was created by Congress to lead "a broad scientific effort to understand the nature of aging and to extend the healthy, active years of life," lending legitimacy—and providing federal funding—to biogerontology researchers across the country.

Most antiaging research in recent years has focused on genetic manipulation and its potential to delay the biological

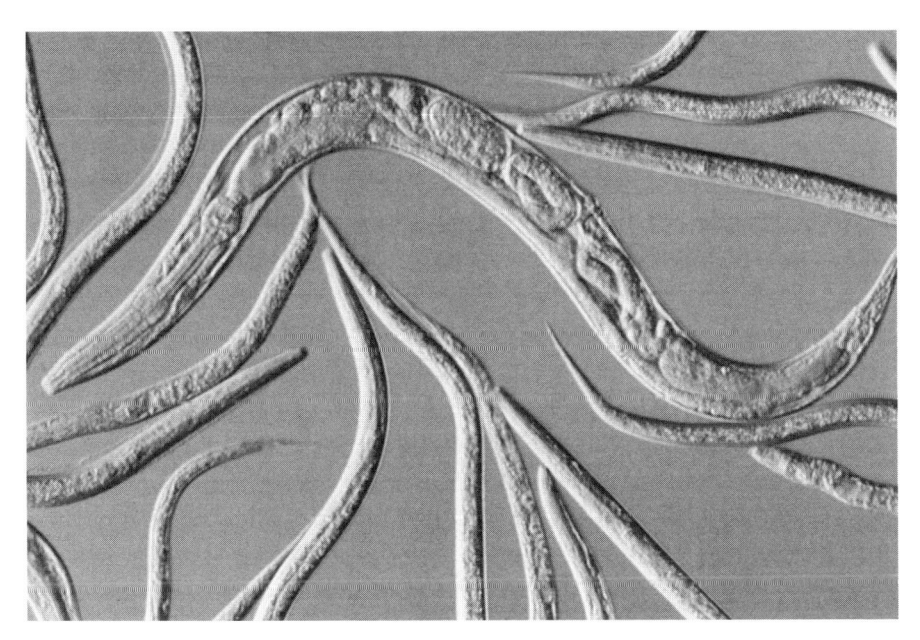

In 1993, researchers at UCSF discovered that a single change in the DAF2 (decay accelerating factor) gene doubled the life span of the tiny roundworm, *C. elegans*. *(Sinclair Stammers/Photo Researchers, Inc.)*

mechanisms of aging. In 1993, biochemist Cynthia Kenyon and her team of researchers at the University of California at San Francisco (UCSF) discovered that a single change in the *DAF2* (decay accelerating factor) gene doubled the life span of the roundworm, *C. elegans*. Normally the worm lives 21 days, but with this single genetic change it lives for 45 days. Dr. Kenyon likened her research to looking at a 90-year-old person who is perfectly indistinguishable from a 45-year-old.

In June 2003, UCSF researchers mapped the genetic changes that resulted from the single manipulation of the DAF2 gene and found that it delays the effects of aging in several ways: It fights infection, changes metabolism, and controls the cellular stress response, while also quieting other genes that work to shorten the life span. "This study tells us," said Dr. Kenyon, "that there are many genes that affect lifespan, each on its own having only a small effect. The beauty of the DAF2 gene is that it can bring all of these genes together into a common regulatory circuit. This allows it to produce these enormous effects on lifespan."

The DAF2 manipulation not only delays death but also the onset of age-related diseases like *Huntington's disease* and cancer in roundworms and mice. Dr. Kenyon and other researchers hope that the same type of process will work to extend the human life span—and the human "health span."

"The consequences are stunning," Dr. Kenyon said in a National Institutes of Health profile of her work, "and if we can figure out a way to copy these effects in humans, we might all be able to live very healthy long lives." But not everyone agrees that this is an appropriate use of scientific resources. In March 1999, the *New York Times* recounted an incident occurring after Dr. Judith Campisi, who studies cellular and molecular mechanisms of aging at Berkeley National Laboratory, gave a public talk on the subject. A number of opponents of prolongevity approached her to say, "How dare you do this research?" They cited extreme environmental degradation by humans, even given our current life span.

"I was really quite taken aback," said Dr. Campisi. "It was a small group but they just about nailed me to the wall."

ETHICAL POSITIONS ON EXTENDING THE HUMAN LIFE SPAN

Christine Overall, a feminist philosopher with Queen's University in Canada, has said in her article "Longevity, Identity, and Moral Character," which appears in the book *The Fountain of Youth: Cultural, Scientific, and Ethical Perspectives on a Biomedical Goal*, that "other things being equal, a longer life is a better one, provided that one is in a minimally good state of health." The case for longer life, she says, "is founded on a genuine appreciation of human potential, of what people want in their lives and are capable of doing and experiencing when given more opportunities."

Most people would agree that increased longevity (when combined with some standard of health) is a good thing, but others are not convinced. Prominent physician and medical ethicist Leon Kass in his book *Toward a More Natural Science* says, "To know and to feel that one goes around only once, and that the deadline is not out of sight, is for many people the necessary spur to the pursuit of something worthwhile . . . Mortality makes life matter." Not every life-extending project in medicine is worthwhile, he argues, and the acceptance of the inevitable end can prepare people for death and enhance their appreciation for life.

A different criticism of prolongevity comes from theologian Audrey Chapman, former director of the Science and Human Rights Program of the American Association for the Advancement of Science. In her article, "The Social and Justice Implications of Extending the Human Life Span" (also in *The Fountain of Youth*), Chapman asks whether it is fair to devote scientific resources to investigating the biological process of aging when many people lack access to basic medical needs. "Enabling a minority of wealthy individuals to secure long life for themselves," she says, "at the expense of further endangering the environment and depriving the poor of resources needed for development" would be fundamentally unjust.

Supporters of biogerontology research counter that it is unfair to single out antiaging research projects when other spending choices made by wealthy nations contribute much more to global inequality, and when the central goal of less controversial medical research (into cancer and heart disease, for example) is also to extend the healthy life span. Nick Bostrom, philosopher and director of Oxford University's Future of Humanity Institute, contends, "For any possible problem that might arise, one question that we must not fail to ask ourselves is: 'Is this problem *so* bad that it is worth sacrificing up to 100,000 lives per day to avoid having to solve it?'"

LIFE EXPECTANCY AT THE START OF THE 21ST CENTURY

Death, on average, is getting farther away for Americans and for people of most other countries. In 1901, life expectancy in the United States was 49; by the close of the century, it had risen to 77. That is an "extension" on life of more than 50 percent. Similar increases have been seen globally, though improvements have been most dramatic in wealthy countries. The single greatest factor responsible for increased average life spans is the sharp decrease in infant mortality, with other dietary, public health, and medical improvements also contributing to longer lives.

In AIDS-stricken countries, primarily in sub-Saharan Africa, the picture is bleaker. The epidemic has slashed life expectancy in some of these countries by more than 30 years. In 2004, the life expectancy in South Africa was 48, in Swaziland it was 37, and in Zimbabwe it was 36. Other countries left out of the global old-age boom are war-torn nations like Iraq, where life expectancy in 2004 was 55; Afghanistan, where it was 42; and Sudan, where it was 58. Former Soviet republics have also suffered a sharp decline in life expectancy since the fall of the Soviet Union. In Russia, for example, life expectancy for men in 2004 was 59, which sadly was below the legal retirement age of 60.

Disparities in life expectancy are not just seen across borders, but within them. In 2006, a team of researchers at the Harvard School of Public Health looked at gaps in life expectancy across different groups of Americans using a combination of racial and geographic categories they termed the "Eight Americas." The results were startling: An African-American man living in a high-risk American city had a life expectancy closer to people in West African countries than to the average white American (based on 2001 data). He could expect to live 21 years less on average than an Asian-American woman. There were other, less drastic disparities: A white farmer from the Great Plains could expect to live about a year longer than a white Middle American and about four years longer than a white farmer from Appalachia or the Mississippi Valley. A member of what researchers termed "Black Middle America" could expect to live about five years less than a member of "Middle America"—consisting mostly of urban and suburban whites. Alarmingly, the gaps were similar to what they had been in 1987.

The "Eight Americas" highlighted in the study are as follows:

- Asian Americans from all over the country
- low-income rural whites in the Great Plains
- Middle America, which is 98 percent white
- low-income whites in Appalachia and the Mississippi Valley
- Western Native Americans
- low-income rural blacks in the Mississippi Valley and the Deep South
- blacks in high-risk urban environments
- the rest of black America, or "Black Middle America"

Perhaps the most revolutionary findings of the study, published in the September 2006 issue of the online journal *PLoS Medicine*, involve not who dies or when, but how and how not. When homicide and AIDS are ignored, the high mortality rate for urban blacks persists. Major causes of death in that group include heart disease, stroke, diabetes, *cirrhosis*, and fatal injuries. The study also debunked the myth that Asians lose their "survival advantage" when they move to the United States and change their eating habits and lifestyle. Asian women in the United States, the researchers found, live an average of three years longer than Japanese women—the longest-lived national group in the world.

Discrepancies in life expectancy cannot be explained by gaps in infant mortality, researchers said, even though the newborn death rate in the United States did make international headlines in 2006 by tying for the second worst in the industrialized world. That finding, announced in the annual *State of the World's Mothers* report, was attributed largely to high infant death rates among minorities and other disadvantaged groups and to a lack of access to basic health care. Inequalities in infant mortality were present in the Harvard study as well, but the most pronounced differences were not among children or the elderly. "The mortality disparities," researchers noted, "are most concentrated in young and middle-aged males and females, and are a result of a number of chronic diseases and injuries with well-established risk factors."

"The magnitude of the life expectancy disparity is most striking and is perhaps a bit larger than I might have guessed," UCLA physician Mitchell Wong, who has looked at how certain diseases contribute to discrepancies in mortality across racial and ethnic groups, told the *Washington Post* in September 2006. "However, it is not surprising that by combining race and geography, disparities are even larger."

Despite stark health discrepancies within and across borders, the United Nations has predicted that by 2050, the world's elderly population (aged 60 and over) will outnumber the world's youth (aged 14 and under) for the first time in human history. An estimated 10 percent of Africans, 25 percent of Asians, 35 percent of Europeans, 20 percent of Latin Americans, 30 percent of North Americans, and 25 percent of Oceanians will be at least 60 years of age, and there will be a projected 2.2 million *centenarians* (people at least 100 years old) by 2050, up from 135,000 in 1998.

LONG-TERM CARE

Whether or not major strides are made in antiaging research, people are already living longer lives, and the demands for long-term care—both home- and community-based—are increasing. According to the Department of Health and Human Services, in 2007 about 9 million men and women over the age of 65 needed long-term care. By 2020, this number is expected to rise to 12 million. Most people will be cared for at home, and according

(continues on page 132)

THE GROWING GLOBAL BURDEN OF CHRONIC DISEASE

Certain noninfectious diseases like type 2 diabetes, obesity, cancer, stroke, and heart disease traditionally have been associated with life in the wealthiest societies, since risk factors for these diseases include increased automobile use and less exercise; more high-fat, high-sugar, processed foods; increased use of tobacco and alcohol; widespread availability of antibiotics and vaccines; and longer life spans. The idea that wealthy countries suffer more from these diseases, however, has come under increasing fire as public health researchers find that *chronic diseases* are becoming the leading killers in many low- and middle-income countries.

According to the World Health Organization (WHO), in almost all countries it is the poorest people who are most at risk for developing chronic health conditions like heart disease, stroke, and diabetes, and for dying prematurely from them. Poor people are most vulnerable because they are exposed to more risk factors without access to adequate health screening and treatment services. The WHO warns that chronic diseases create and deepen poverty, as people who are disabled by them often cannot continue to work or provide care for other family members.

Why are chronic diseases hitting low- and middle-income countries harder than ever? Many factors seem to be at play and are the focus of intense inquiry in the public-health field. As cities grow and as people assume more of an urban lifestyle, risk factors such as a sedentary lifestyle and addiction to nicotine increase. Globalization and the availability of fast food have meant a shift toward "Western" diets high in saturated fats and sugar, and dramatic increases in life expectancy in most parts of the world have meant that age-related diseases like cancer are on the rise. Thomas Novotny, Director of International Programs at the UCSF School of Medicine, argued in his 2005 *PloS Medicine* article, "Why We Need to

(continues)

(continued)

Rethink the Diseases of Affluence," that more detailed analysis of how economic development is associated with risk factors for noninfectious diseases "might dispel prejudices about the 'diseases of affluence' and stimulate policy approaches and research that appropriately target emerging risk groups across the globe, regardless of socioeconomic status."

Dr. Hiroshi Nakajima, former Director-General of the WHO, said in the organization's 1997 report, *Conquering Suffering,*

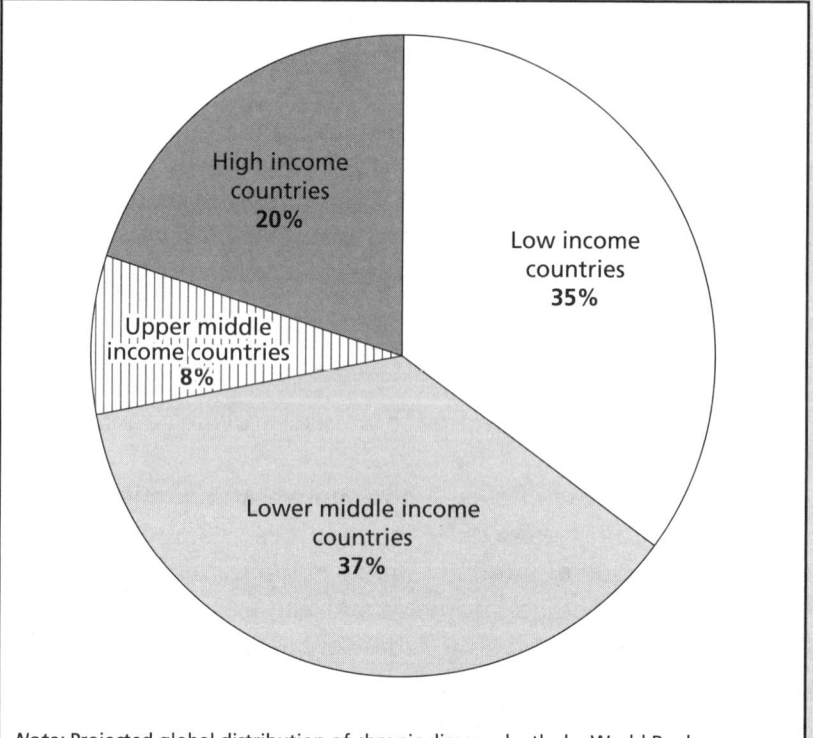

Note: Projected global distribution of chronic disease deaths by World Bank income group, all ages, 2005

© Infobase Publishing

Projected global distribution of chronic disease deaths by World Bank income group, all ages, 2005 *(Source: World Health Organization, "Preventing Chronic Diseases: A Vital Investment," 2005)*

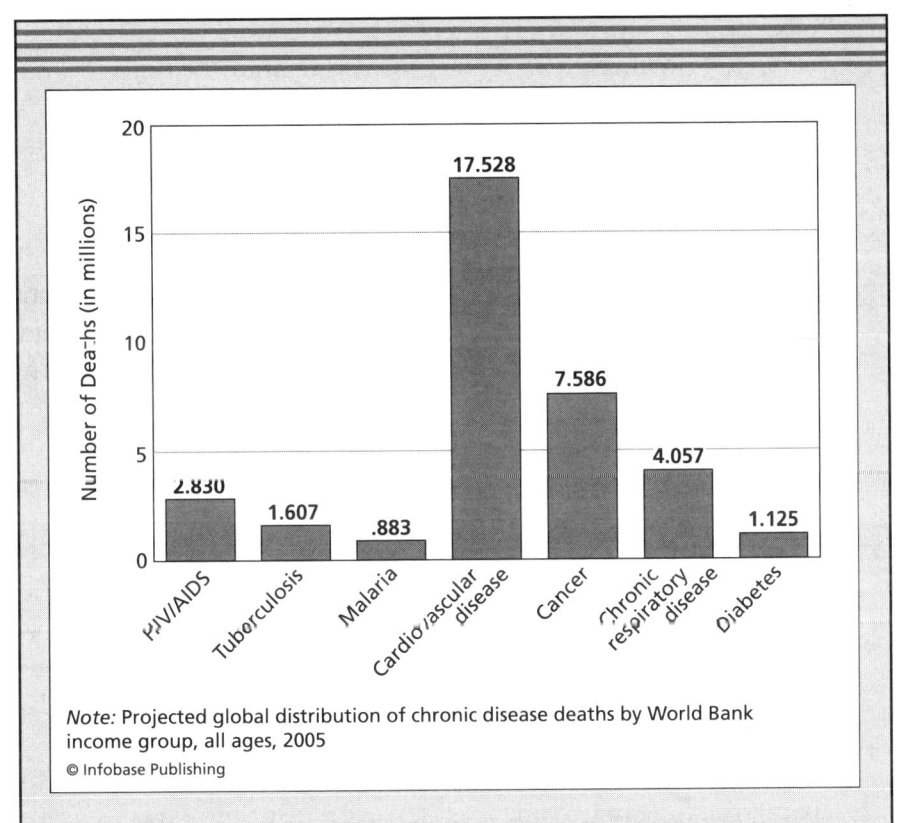

Life Extension, Aging, and Palliative Care

[Bar chart showing number of deaths (in millions):
- HIV/AIDS: 2.830
- Tuberculosis: 1.607
- Malaria: .883
- Cardiovascular disease: 17.528
- Cancer: 7.586
- Chronic respiratory disease: 4.057
- Diabetes: 1.125]

Note: Projected global distribution of chronic disease deaths by World Bank income group, all ages, 2005
© Infobase Publishing

Projected global deaths by cause, all ages, 2005 *(Source: World Health Organization, "Preventing Chronic Diseases: A Vital Investment," 2005)*

Enriching Humanity, "We dare not turn our back on infectious diseases.... But neither can we ignore the growing burden in suffering and disability represented by noncommunicable diseases...." Public health research points to some of the keys to alleviating the chronic disease burden, like better food, tobacco, and alcohol policies; improved urban planning and environmental engineering; and widespread disease screening and availability of affordable generic medicines and nicotine replacements. "[I]n celebrating our extra years," said Dr. Nakajima, "we must recognize that increased longevity without quality of life is an empty prize ... [that] *health* expectancy is more important than *life* expectancy."

(continued from page 128)
to AARP (formerly the American Association of Retired Persons) and the National Alliance for Caregiving, about 44 million Americans, or an estimated 21 percent of U.S. households, already provide unpaid care to an adult relative or friend.

The 2005 AARP report, *Beyond 50: A Report to the Nation on Independent Living and Disability,* stated that among those 50 and older with a disability who had regular help with everyday activities like bathing and cooking, 75 percent received unpaid help from family members (usually a spouse or a child). Nearly three out of 10 respondents said they needed more help than they were receiving, and nearly three-quarters said the main reason they did not have enough help was that they could not afford it.

Nursing-home care remains the most expensive long-term care option. The average cost for one year in a nursing home rose above $70,000 in 2006—a cost Medicare does not cover. More than 40 percent of elderly Americans will likely need to enter a nursing home at some point in their lives, and about 10 percent of those will need to stay for five years or more.

Whatever long-term care people choose (or have available to them), the time may come for intensive end-of-life care. The *hospice* care movement is grounded in the need for humane, holistic treatment for an increasingly older and debilitated population as its members reach the end of life.

HOSPICE AND PALLIATIVE CARE

The word "hospice" is rooted in the Latin *hospitium,* or guesthouse. Hospices started many centuries ago as places of rest for tired or ill travelers. The first modern hospice was St. Christopher's, founded in 1967 in South London by Cicely Saunders. Saunders, a prominent nurse, physician, social worker, and writer, believed that dying patients and their families should have access to care and support from an interdisciplinary team of caregivers, and that patients' suffering should be alleviated with modern pain-management techniques.

Dr. Elisabeth Kübler-Ross is credited with planting seeds of the hospice care movement in the United States. A practicing

The first modern hospice, St. Christopher's, was founded in 1967 in South London by Cicely Saunders (1918–2005), a prominent nurse, physician, social worker, and writer. Dame Saunders is pictured here after being awarded the world's largest humanitarian award, the Conrad N. Hilton Humanitarian Prize, in August 2001. *(Adrian Dennis/AFP/Getty Images. Hospice photo courtesy of St. Christopher's Hospice)*

psychiatrist, Kübler-Ross was struck by the often-callous treatment of dying patients in hospitals. She established a seminar at the University of Chicago to educate graduate students about the dying process, and to expose them to terminally ill patients and their stories. Her groundbreaking book, *On Death and Dying*, was published in 1969.

In 1974, the first modern hospice in the United States was founded in New Haven, Connecticut. Since that time the hospice movement has grown dramatically, with more than 3,200 hospice programs caring for almost 885,000 dying people in 2002.

Hospice care typically is available to patients living the last six months of life, for whom aggressive treatments aimed at curing their disease are no longer appropriate. Eighty percent of hospice care is provided in patients' homes, loved ones' homes, and nursing homes, with some inpatient hospice services available in a hospital setting.

Conditions treated by hospice include cancer, heart disease, pulmonary disease, and AIDS, as well as Alzheimer's disease and other neurological disorders. Teams of doctors, nurses, social workers, spiritual counselors, therapists, and an estimated 400,000 volunteers in the United States alone provide pain and symptom relief to patients, as well as emotional and spiritual support for the feelings of fear and loneliness felt by patients, families, and friends. Primary caregivers learn how to provide the best possible care for their loved ones, and many caregivers say that in the process, they grow more comfortable with their own mortality.

Hospice care is sometimes called palliative care, though palliative care is a broader term that includes all pain-relief measures. From the Latin *palliare,* to cloak, palliative care is any form of treatment aimed at alleviating symptoms rather than providing a cure, and can be used simultaneously with curative treatment. Palliative care may include treatment of side effects of a curative treatment like chemotherapy. Unlike hospice care, palliative care is available to patients in the United States regardless of their diagnosis or prognosis.

SUMMARY

Throughout most of the world, life expectancy has increased during the last century, thanks to rapid advances in preventive care and medical treatment. Wealthy countries like the United States have seen dramatic improvements, but within U.S. borders, gaps in life expectancy between rich and poor and among racially and ethnically diverse groups are significant. A growing effort in public health seeks to understand the causes of these disparities, and how equality of life expectancy—and of "health expectancy"—might be achieved. At the same time, longer lives mean more demand for quality, affordable long-term care from a system already unable to meet the needs of many elderly Americans and their loved ones.

Researchers in the field of biogerontology are hunting for the biological keys to human longevity, and the potential implications of their efforts are exciting and also largely unknown. Dr. Gregory Stock, director of the Program on Medicine, Technology, and Society at the University of California at Los Angeles's School of Public Health, told the *New York Times* in March 1999,

Life Extension, Aging, and Palliative Care

"It is going to be very hard for us to deal with." The idea that we could double the human life span—or more—Dr. Stock said, "puts a distance between ourselves and all of our history . . . All of human wisdom on how to live a life" would no longer be relevant. People's relationships to life and death—and everything in between—might change in countless unforeseen ways.

7

New Technology and the Cost of Treatment

Modern medical science has brought health and hope to millions of people. Its contributions to human welfare and quality of life cannot be overestimated. Previous chapters in this book have highlighted exciting new discoveries and some of the ethical challenges they pose, but there is one issue yet to consider: The high price tag attached to research, development, and high-tech treatment that has contributed to the staggering rise in health-care costs in recent years and has exacerbated the challenge of providing a decent minimum of care for everyone.

This chapter considers some of the causes and consequences of rising medical costs, and takes a closer look at the example of the HIV/AIDS epidemic: The amazing scientific discoveries that have extended millions of lives, the extreme costs of treating the disease, and current efforts to narrow the survival gap between wealthy and lower-income nations.

THE RISING COSTS OF HEALTH CARE

In 2006, national health-care expenditures climbed 6.7 percent to an astronomical $2.1 trillion, or $7,026 per person. Medical costs represented 16 percent of the gross domestic product

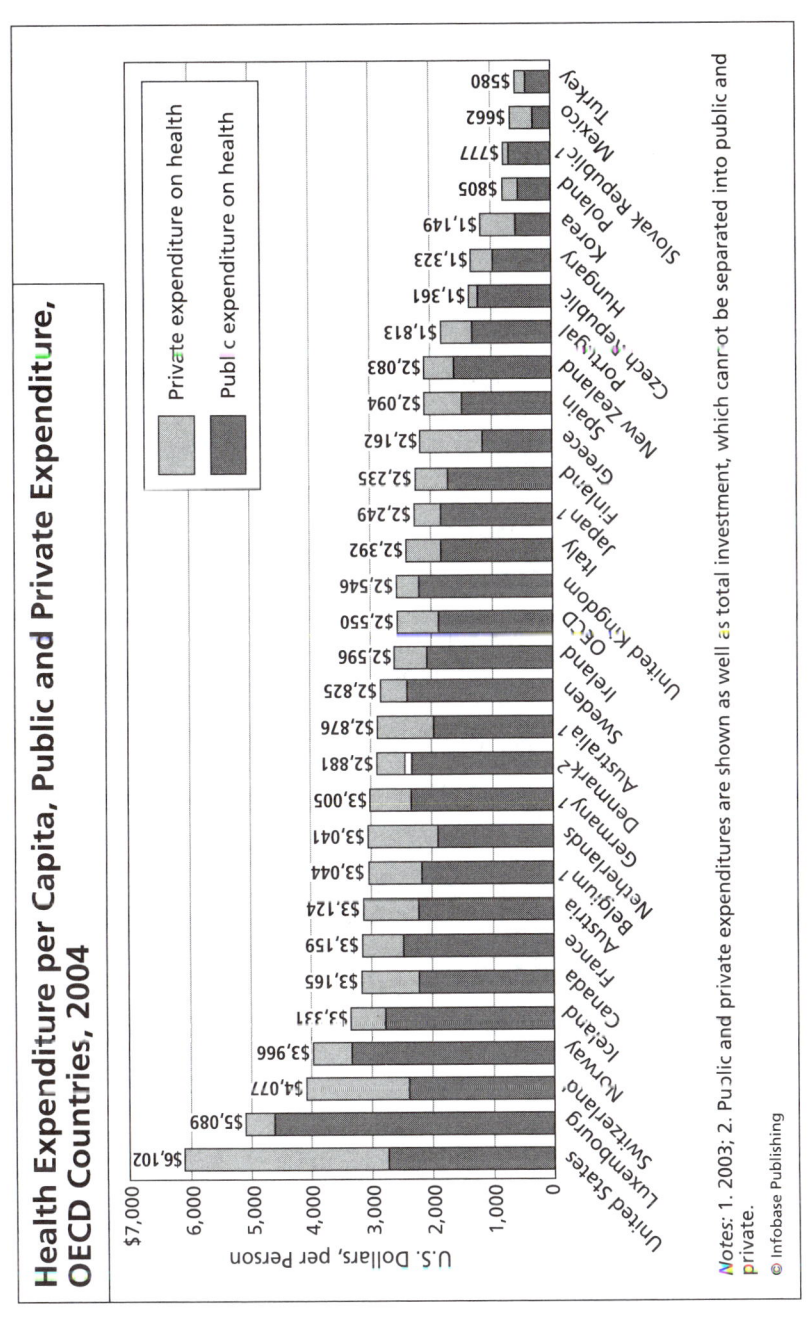

New Technology and the Cost of Treatment

Health Expenditure per Capita, Public and Private Expenditure, OECD Countries, 2004

Legend: Private expenditure on health; Public expenditure on health

Country	U.S. Dollars, per Person
United States	$6,102
Luxembourg	$5,089
Switzerland	$4,077
Norway	$3,966
Iceland	$3,331
Canada	$3,165
France	$3,159
Austria	$3,124
Belgium 1	$3,044
Netherlands	$3,041
Germany 1	$3,005
Denmark 2	$2,881
Australia 1	$2,876
Sweden	$2,825
Ireland	$2,596
United Kingdom	$2,550
OECD	$2,546
Italy	$2,392
Japan 1	$2,249
Finland	$2,235
Greece	$2,162
Spain	$2,094
New Zealand	$2,083
Portugal	$1,813
Czech Republic	$1,361
Hungary	$1,323
Korea	$1,149
Poland	$805
Slovak Republic 1	$777
Mexico	$662
Turkey	$580

Notes: 1. 2003; 2. Public and private expenditures are shown as well as total investment, which cannot be separated into public and private.

© Infobase Publishing

Health expenditure per capita in OECD countries, public and private expenditure, 2004 *(Source: OECD Health Data, June 2006)*

(*GDP*), a higher percentage than any other industrialized nation, and nearly twice that of the United Kingdom, which provides *universal health care* to its citizens (see the following table). Health care spending in the United States is expected to continue to grow, reaching a whopping $4 trillion, or 20 percent of GDP, by 2015.

High costs do not always correlate with quality. A 2007 study by the Commonwealth Fund compared the United States and five other countries—Australia, Canada, Germany, New Zealand, and the United Kingdom—all of which provide universal health care for their citizens. The study found that the U.S. system ranks last or next-to-last on five indicators of success: quality, access, efficiency, equity, and healthy lives. The United States is strong on preventive care and access to specialists, according to the report, but "weak on access to needed services and ability to obtain prompt attention from physicians."

Rising health-care costs have become a hardship for millions of Americans. The annual premium for an employer-based health plan (for a family of four) averaged $12,100 in 2007, and workers contributed nearly $3,300 to this number—10 percent more than in 2006. Annual premiums for family cover-

HEALTH CARE SPENDING IN SEVEN INDUSTRIALIZED COUNTRIES

Country	Health Costs as Percentage of Gross Domestic Product (GDP)
Canada	9.9
France	10.5
New Zealand	8.4
Sweden	9.1
Switzerland	11.6
United Kingdom	8.3
United States	15.3

Source: OECD Health Data 2006 (June 26, 2006)

age exceeded the gross earnings for a full-time, minimum-wage worker ($10,712).

A 2005 report by the Kaiser Family Foundation found that premiums for employer-sponsored health insurance in the United States had risen five times faster since 2000, on average, than salaries earned by workers. Almost 50 percent of Americans are very worried about having to pay more for health care or insurance, says the National Coalition on Health Care (NCHC), while 42 percent are afraid that they will not be able to afford medical care they need. "Retiring elderly couples will need $200,000 in savings just to pay for the most basic medical coverage," says NCHC. "Many experts believe that this figure is conservative and that $300,000 may be a more realistic number."

HEALTH CARE COVERAGE IN THE UNITED STATES

When health care costs rise, so do the costs of private health insurance. According to the Census Bureau's 2005 survey, 45.8 million Americans (roughly 15 percent of the population)—including about 9.6 million children—currently go without medical insurance.

People without medical insurance are more likely to be young: Of the uninsured, 21 percent are under 18 and 63 percent are under 34. The uninsured are also disproportionately Hispanic and to a lesser degree African American. Thirty percent of people without medical insurance are Hispanic, whereas Hispanics represent about 14 percent of the total population. Fifteen percent of people without medical insurance are African American, though African Americans represent about 12 percent of the total population.

The news is not all bad: The number of uninsured children decreased by almost 2 million between 1998 and 2003—thanks in part to widespread efforts to enroll children either in Medicaid (federally assisted health care) or in the State Children's Health Insurance Program (SCHIP). Yet millions more children (about seven out of ten without insurance) are eligible for low-cost or free health care but are not yet enrolled.

At a February 2007 meeting of the National Governors Association, 14 states announced that they were running out of federal money for the Children's Health Insurance Program and many governors were concerned that without additional federal money, their states would be forced to enroll fewer children. Governor Jim Douglas, Republican of Vermont, told the *New York Times* in February that his state had made tremendous progress in covering children (less than 4 percent of children were without insurance) and that "we don't want to lose ground." Governor Jon Corzine, Democrat of New Jersey, said that states want to avoid "rationing health care to our most vulnerable and our most needy."

In October and December 2007, President George W. Bush vetoed two children's health bills that would have provided an additional $35 billion for the program, bringing total funding to $60 billion over five years. The bills were designed to continue coverage for the 6.6 million children already insured under the program, while adding nearly 4 million children to the rolls. In lieu of long-term funding, Congress and the president approved a short-term budget extension to cover enrolled children for an additional year.

Not surprisingly, a lack of insurance coverage correlates with poor health care for children. In 2003, a startling one-third of all uninsured children went without any medical care for the entire year, whereas nearly 88 percent of children with medical insurance received at least some care. And when children become sick, the situation is even more serious: Uninsured children in America are 10 times more likely to miss out on necessary medical care than children enrolled in some form of medical insurance.

COSTS OF NEW TECHNOLOGY

What is driving up health care costs so astronomically? Some factors lie outside the scope of this chapter, including the healthcare needs of an aging population (see chapter 6), administrative waste in a mixed public and private system, and the cost of malpractice insurance. This section will focus primarily on the

New Technology and the Cost of Treatment

Considered Cost-Effective		Considered Less Cost-Effective	
Choice of treatment	Cost per quality-adjusted life year*	Choice of treatment	Cost per quality-adjusted life year*
Flu vaccinations for newborns v. none	Less than $0; costs less to vaccinate than to treat later in life	Annual screening for depression v. none	$100,000 to $500,000
Yearly mammogram v. one every two years	$10,000 to $25,000	M.R.I. for high-risk children with headaches v. no treatment	$100,000 to $500,000
High blood pressure medication v. none	$10,000 to $60,000	Heart pump for patients with end-stage heart failure v. no device	$500,000 to $1.4 million

Example: 1,000 patients are given hypertension medication at a total cost of $1 million. The treatment is expected to prevent one fatal and one nonfatal heart attack. This saves 30 years of full health—25 from the prevented fatal attack and 5 from the nonfatal. The total cost is divided by 30 years, resulting in roughly $33,000 for each quality-adjusted life year.

*The net costs to society for each year of healthy life gained.
© Infobase Publishing

Determining the cost-effectiveness of medical treatment choices *(Adapted from the* New York Times, *"What Is Good Health Worth?" June 8, 2007. Original source: Center for the Evaluation of Value and Risk in Health, Tufts University and New England Medical Center)*

extraordinarily high costs associated with research, development, and delivery of new drugs and medical technologies.

Medical ethicist Daniel Callahan, in his book *False Hopes*, notes that during a period when per capita spending for health care increased by 269 percent, per capita health-care technology costs increased by 791 percent. Much of the issue can be attributed to the availability of expensive technologies (MRI units, for example), not just in regional medical centers as they are in many other industrialized countries—but in local clinics and physicians' offices as well. Specialists must be trained to operate complex new technologies, and are paid higher fees for their expertise.

The rising cost of prescription drugs is another major culprit in research and development (R & D) costs. The average cost to bring a single medicine to market has been estimated at 16 years and more than $800 million. Included in this price tag are any marketing and legal costs of promoting drugs to doctors and patients or protecting patent rights.

In 2002, the *Wall Street Journal* broke the story of the Shark Fin Project—an effort by the pharmaceutical company AstraZeneca to extend patent rights to their lucrative heartburn medication Prilosec. It was by no means an isolated case within the industry, but it became a lightening rod for frustrations over misplaced priorities. The Shark Fin Project, the *Journal* reported, settled on the idea to tweak Prilosec—which was composed of a left-hand and right-hand version of the same molecule—to effectively cut it in half. The result was the drug Nexium, which received a new patent from the U.S. Patent Office and became available to consumers at the cost of $120 for a one-month supply.

"The Prilosec pattern," said the *Wall Street Journal,* "repeated across the pharmaceutical industry, goes a long way to explain why the nation's prescription drug bill is rising an estimated 17% a year even as general inflation is quiescent."

The responsibility of lowering drug costs lies not only with pharmaceutical companies, but with doctors as well, who sometimes prescribe more expensive drugs without clear proof that they are more effective. Part of the problem, says John Abramson, instructor of clinical medicine with Harvard Medical School, is that busy doctors often rely upon dubious findings published in trusted medical journals when deciding which drugs to prescribe for their patients.

In his book *Overdosed America,* Abramson highlights a study of Pravachol—a drug prescribed for heart disease—and the misleading information about it that appeared in a prestigious medical journal. The article suggested that Pravachol helped to prevent strokes in heart disease patients, based on a study con-

New Technology and the Cost of Treatment

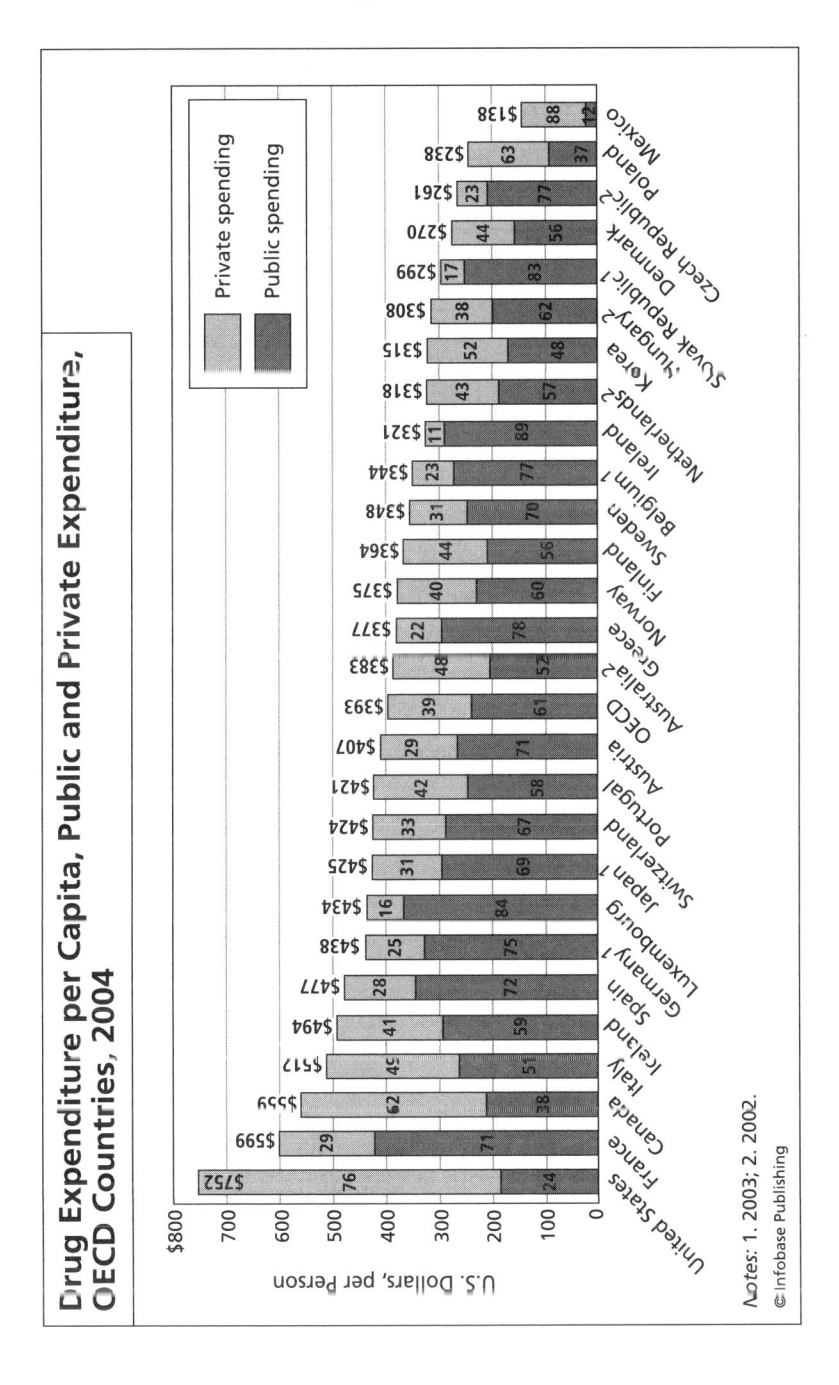

Drug Expenditure per Capita, Public and Private Expenditure, OECD Countries, 2004

Country	Public	Private	Total
United States	24	76	$752
France	71	29	$599
Canada	38	62	$559
Italy	51	45	$517
Iceland	59	41	$494
Spain	72	28	$477
Germany[1]	75	25	$438
Luxembourg	84	16	$434
Japan[1]	69	31	$425
Switzerland	67	33	$424
Portugal	58	42	$421
Austria	71	29	$407
OECD	61	39	$393
Australia[2]	52	48	$383
Greece	78	22	$377
Norway	60	40	$375
Finland	56	44	$364
Sweden	70	31	$348
Belgium[1]	77	23	$344
Ireland	89	11	$321
Netherlands[2]	57	43	$318
Korea	48	52	$315
Slovak Republic[2]	62	38	$308
Hungary[2]	83	17	$299
Czech Republic[2]	56	44	$270
Denmark	77	23	$261
Poland	37	63	$238
Mexico	12	88	$138

Notes: 1. 2003; 2. 2002.
© Infobase Publishing

Drug expenditure per capita in OECD countries, public and private expenditure, 2004 *(Source: OECD Health Data, June 2006)*

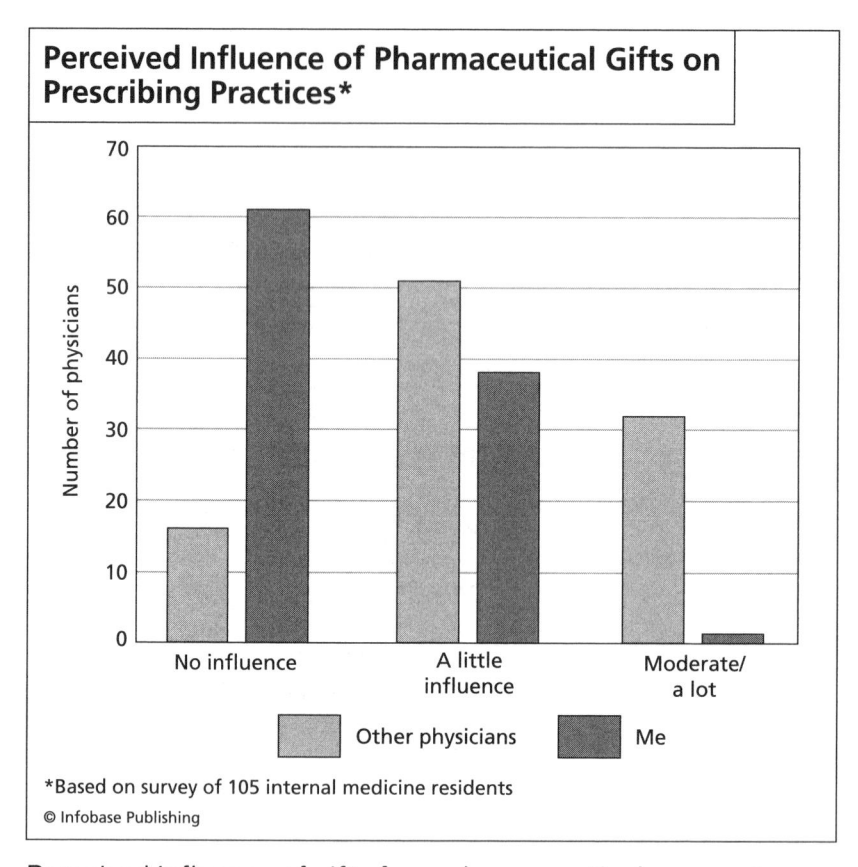

Perceived Influence of Pharmaceutical Gifts on Prescribing Practices*

*Based on survey of 105 internal medicine residents
© Infobase Publishing

Perceived influence of gifts from pharmaceutical companies on physicians' prescribing practices *(Adapted from M. Steinman et al., "Of Principles and Pens: Attitudes and Practices of Medicine Housestaff toward Pharmaceutical Industry Promotions,"* The American Journal of Medicine *110, no. 7 [2001]: 551–557)*

ducted with patients whose average age was 62. For every thousand patients on Pravachol, one stroke was averted (at a cost of at least $1.2 million for each prevented stroke). In patients over 70, however—an age group much more likely to suffer from stroke—more strokes occurred with the medication than with a placebo.

"When I finished analyzing the article," says Abramson, "and understood that the title didn't tell the whole story, that the findings were not statistically significant, and that Pravachol appeared to cause *more* strokes in the population at greater risk, it felt like a violation of the trust that doctors (including me) place in the research published in respected medical journals." The line between successful product marketing and meaningful scientific research has become too blurred in some cases for health professionals to discern the truth without a detailed reading.

Marcia Angell, former editor-in-chief of the *New England Journal of Medicine,* argues in her 2004 book, *The Truth about the Drug Companies,* that instead of "investing more in innovative drugs and moderating prices, drug companies are pouring money into marketing, legal maneuvers to extend patent rights, and government lobbying to prevent any form of price regulation." The public's interests would be much better served, Angell contends, if drug companies were to take the money they spend on marketing and use it to guarantee life-saving drugs at affordable prices.

There is new reason to hope that critics of for-profit drug companies may one day get their wish. An entirely new American phenomenon—the nonprofit pharmaceutical company—is enjoying the spotlight after a fledgling company, Institute for OneWorld Health, conducted a successful clinical trial in India in 2004 (thanks in part to a $47.25 million grant from the Bill and Melinda Gates Foundation). The trial showed that an old drug called paromomycin—first developed in the 1950s and marketed for the treatment of intestinal parasites—could cheaply and effectively treat the deadly disease visceral leishmaniasis, or *VL* (also known as *black fever*). The drug could save an estimated 200,000 lives a year.

The new nonprofit has also formed a partnership with the Berkeley developer of an inexpensive process to synthesize an important malarial drug—a drug previously unavailable to many citizens of low-income countries due to its excessive production costs. *Malaria* causes an estimated 1 million to 3 million deaths each year worldwide, and it is the leading killer of children under the age of five in many countries.

The next section will take a detailed look at the global fight against another deadly disease, HIV/AIDS: The elusive nature of the virus, the high cost of the drugs used to treat it, and the funding strategies that have worked best to save lives.

THE HIV/AIDS PANDEMIC: DISPARITIES IN DISTRIBUTION OF CARE

During the last 25 years, some 30 million people have died of AIDS and some 40 million more have contracted the HIV virus. HIV/AIDS continues to spread rapidly across the globe, especially in low-income nations where prevention and treatment efforts have been severely underfunded from the onset of the epidemic. A projected 40 million more people will contract HIV worldwide during the next 10 years.

In the mid-1990s, the discovery of the *triple cocktail*—a combination of drug therapies that exhibited miraculous, life-extending properties—meant that HIV no longer was a death sentence. Not everyone benefited from the life-saving discovery, however. Citizens of low-income nations died from AIDS faster than citizens of wealthier countries. The drugs that kept the virus at bay were expensive, and as of the mid-1990s, the world was spending only $300 million per year (less than 4 percent of current expenditures) on HIV-related programs throughout the entire developing world.

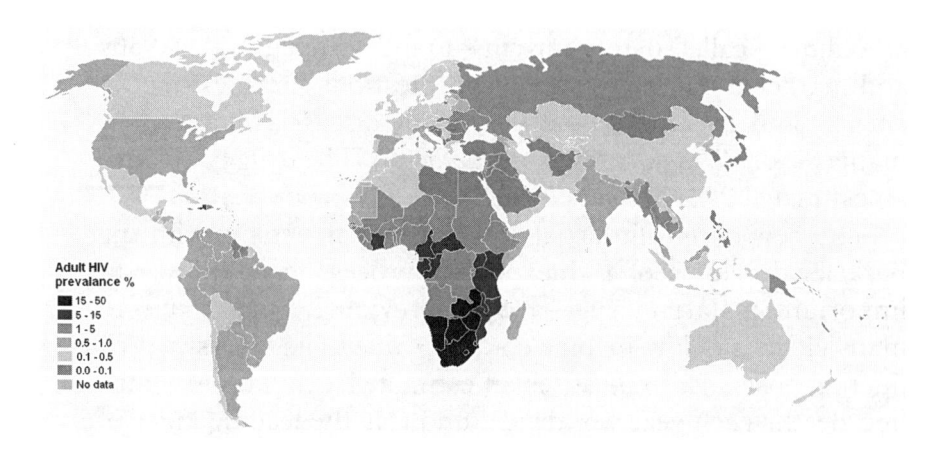

Global distribution of infection in the HIV/AIDS pandemic *(Wikimedia Commons/Grcampbell)*

The financial outlook has improved since then. In 2005, the world pledged $8.3 billion toward HIV/AIDS programs through a combination of international, national, and private efforts (like those of the Bill and Melinda Gates Foundation). This meant an estimated 250,000 to 350,000 fewer deaths in 2005 alone.

There is continuing cause for urgency, however. In 2006, antiretroviral drugs reached only about 20 percent of the people who needed them, and less than one in five high-risk individuals had access to science-based prevention services. Less than 10 percent of families with children orphaned or otherwise made vulnerable by the epidemic had access to care and support services. According to the United Nations Joint Programme on HIV/AIDS (UNAIDS), annual spending of $22.1 billion (nearly triple 2005 expenditures) is needed if there is to be any hope of reversing the epidemic.

Most HIV researchers and activists agree that strong, evidence-based prevention programs are a key to stemming the tide. Says Dr. David Ho, *Time* magazine's 1996 "Man of the Year" for his work on the "triple cocktail" treatment, "We cannot continue just to treat patients as they become infected. The real solution to this epidemic is to curtail the spread of the virus."

The HIV Virus and Its Treatment

The prevailing theory about the origin of HIV in humans is that most cases spread from a single transmission from chimpanzee to human. *HIV-1 M,* the most common and virulent form of HIV, was traced in May 2006 to chimpanzee populations in southern Cameroon, a country in central Africa. A similar route is suspected for HIV-1 subtypes N and O, whereas *HIV-2*—a less virulent species occurring in West Africa—has been linked to a monkey called the sooty mangabey.

Most researchers believe that transmission to humans occurred as a result of the "bushmeat trade"—the practice of hunting and butchering chimps and monkeys. The fact that the HIV-1 M virus—responsible for the great majority of infections worldwide—may have spread from a single point of contact speaks both to our social nature as a species and to the remarkable adaptivity of the virus.

HIV can be passed through blood, semen, vaginal fluid, and breast milk. It is not casually transmitted, and it cannot be contracted through sweat, saliva, or tears. It does not live for more than a few minutes outside the body. Hugging a person with HIV does not transmit the disease, which is primarily spread through the following practices:

- sex (vaginal or anal, and in rare cases, oral)
- blood transfusions
- contaminated needles
- mother-to-child contact at birth
- breast-feeding

U.S. FUNDING FOR HIV TREATMENT AND PREVENTION

Since the beginning of the epidemic, the United States has contributed the most bilateral assistance to HIV programs, accounting for 40 to 50 percent of all HIV program funding in low-income countries by the mid-1990s. In 2003, President George W. Bush pledged to increase U.S. spending by $10 billion (for a total of $15 billion over five years). The President's Emergency Plan for AIDS Relief, or PEPFAR, concentrates funding in 15 African nations (and other countries at a reduced level). The benefits to treatment programs have been remarkable: About one out of every three people receiving antiretroviral drugs in poorer and middle-income nations receive those drugs through funding from PEPFAR.

The prevention arm of PEPFAR has been a major source of controversy, however, as one-third of its funding is allotted to abstinence-until-marriage programs. The United States Government Accountability Office (GAO) found that PEPFAR's abstinence requirements have caused wasteful confusion on the ground and have forced cuts in necessary prevention areas, such as mother-to-child transmission programs.

As frightening an epidemic as it has become, HIV is a preventable disease. Safe-sex practices (condoms or abstinence), clean needles for intravenous drug addicts, and a well-screened blood supply could cut the virus off at three major sources. New medical approaches to prevention are always being explored, such as *microbicides* (for women whose partners will not agree to wear condoms) and circumcision, which has been shown to reduce female-to-male transmission on the order of 60 percent and male-to-female by approximately 30 percent.

There is growing consensus among researchers and patient advocates that routine testing for HIV (like *Pap tests* for cervical cancer) would go far to prevent new cases of the disease.

Most public health officials advocate an "abstinence plus" model, which teaches that abstinence outside of marriage is the best way to avoid HIV and other sexually transmitted diseases (*STDs*) as well as unwanted pregnancies, while also teaching alternatives to abstinence such as condom use.

Studies of abstinence-only programs in the United States have shown that, while they may delay sexual activity in teenagers for short periods (12 to 18 months), they also are associated with higher rates of STDs and pregnancy once sexual activity is initiated, since teens are less willing to use condoms or other contraceptives. According to Dr. Chris Beyrer of the Johns Hopkins School of Public Health, "The net effect is almost nothing, and that is because if people know less, they're more vulnerable."

In April 2008, the U.S. House of Representatives approved a bill that would authorize $50 billion over the next five years to support campaigns against AIDS, tuberculosis, and malaria, and would greatly ease requirements that tie HIV-prevention programs to abstinence-only education. The Senate passed its own version of the bill in July.

The CDC estimates that between 25 and 30 percent of people infected with the virus in the United States do not know that they are infected, and thus are unwittingly transmitting it to others. The CDC recommended in the summer of 2006 that HIV testing become standard medical care for Americans aged 13 to 64.

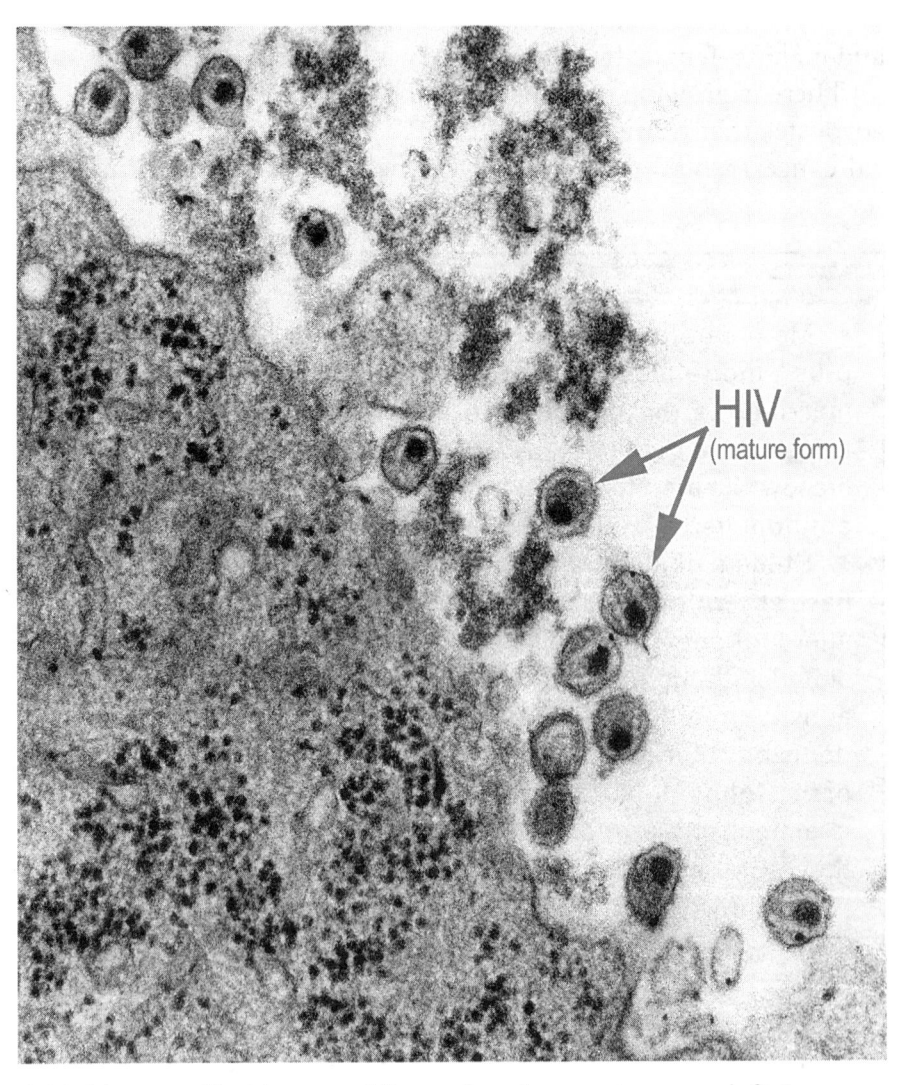

A highly magnified image of the mature human immunodeficiency virus (HIV) *(CDC)*

Other advocates believe that if routine testing becomes the norm, people who are tested should be ensured adequate access to information, support, and treatment. Judith Auerbach, vice president of the American Foundation for AIDS Research, told PBS's *Frontline* in 2006, "People need to know what it means to test positive and to test negative with respect to both treatment and prevention options and responsibilities. Also, it is essential that people are provided access to, not just information about, appropriate care and prevention services once they get their diagnosis." Auerbach points out that HIV medications are still very expensive, and that "people who start taking them must continue with them for the rest of their lives. Many people simply will not be able to afford this."

What about an HIV vaccine? Most researchers believe that a vaccine is years away. The central problem lies with the astonishing power of the virus to harness the body's immune system, turn it against itself, and eventually destroy it. HIV attacks the very control center of disease resistance by attacking the cells that are in charge of our immune response—*CD4 cells,* also called *helper T-cells.* Once inside a CD4 cell, HIV installs its genetic material and uses the cell's own structures to generate more viruses. A vaccine could do more harm than good by boosting the immune system, thereby speeding production of the virus.

To complicate matters further, HIV is a retrovirus—meaning that it is made up of a single-stranded piece of RNA (genetic blueprint) that must be converted to DNA inside a host cell. This process is riddled with mistakes—genetic "typos," if you will. Scientists call these mistakes mutations. HIV's rapid mutation rate makes it especially dangerous, since it quickly finds ways to work around the body's immune response and life-prolonging drugs. This high mutation rate might also increase the chances of a vaccine-resistant virus getting around the immune system and spreading to others.

Scientists know of no one who has been able to rid herself of the virus once it has been contracted. Dr. Anthony Fauci, head of the National Institute of Allergy and Infectious Diseases (NIAID) at the National Institutes of Health (NIH), puts it this way: "Of all the microbes we know of, this is the only one in which the body has proven itself completely incapable of eliminating the virus from the body once it gets infected.... When you develop a vaccine, you look at the body and you say, 'This is what the body did to protect itself, and we're going to develop a vaccine that mimics what the body has done.'" With the HIV virus, researchers do not have that luxury. This is the first time in the centuries-old history of immunology that there has been this enormous barrier to scientific understanding.

In the meantime, there are a host of antiretroviral drugs that prolong life substantially, but do not provide a cure. As of autumn 2008, more than 30 HIV medicines were available. Most of them belong to the following four basic classes:

1. *Reverse transcriptase inhibitors,* which block the action of an *enzyme* that allows HIV's RNA to be converted to DNA. AZT (also known as ZDV), the famous first HIV drug to be discovered, belongs to this family.
2. *Protease inhibitors,* which interfere with virus replication by preventing the breakdown of proteins that allow HIV to reproduce. Unlike reverse transcriptase inhibitors, protease inhibitors can block HIV replication in cells that are already infected. Combined with two reverse transcriptase inhibitors, their benefits can be extraordinary. When this "triple cocktail" treatment was first introduced in 1996, many AIDS patients came back from the brink of death.
3. *Entry inhibitors,* including *fusion inhibitors,* which bind to surface proteins on the CD4 cell or on the HIV virus, thus blocking its ability to fuse with the cell. This breakthrough was thanks to the discovery of a small number of people who were immune to HIV due to a protective abnormality in the surface protein of the CD4 cell (which may also have protected their ancestors from the plague, or "black death," several centuries earlier).
4. *Integrase inhibitors,* which work to stop HIV's genetic material from plugging itself into the host cell's DNA.

New Technology and the Cost of Treatment

Drug therapies can extend life an average of eight years, though some people are able to stay healthy much longer. To maximize benefit, patients must not miss a dose, which means that the drugs must be made consistently available to HIV-positive persons if long-term treatment is to be successful.

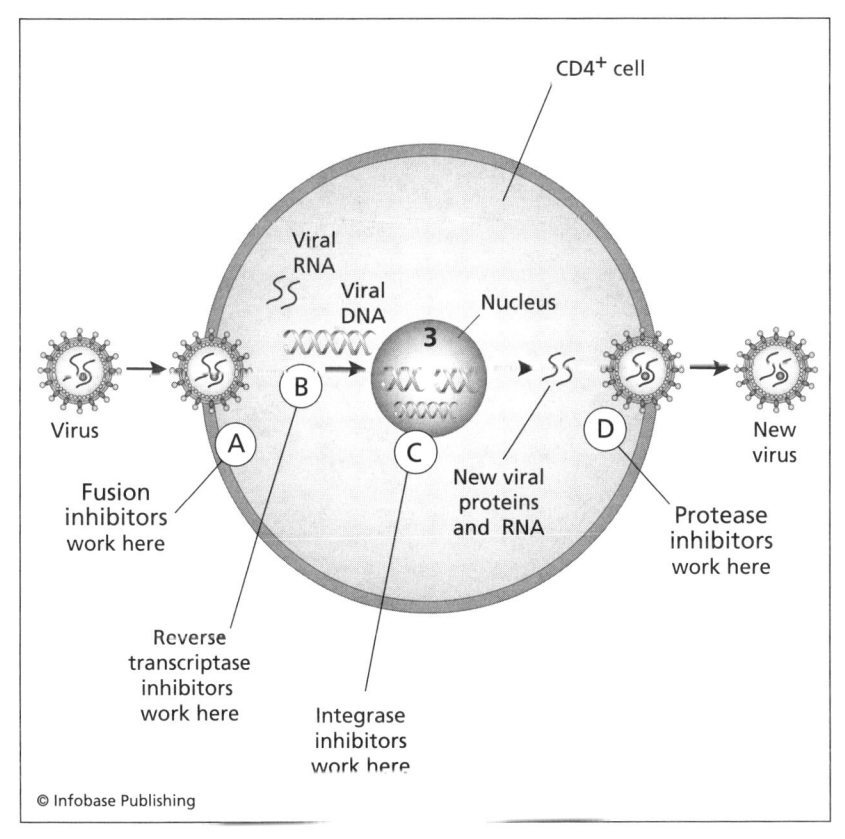

Illustration of how HIV drugs work to interrupt the viral life cycle. (a) Fusion inhibitors block HIV's ability to fuse with the cell. (b) Reverse transcriptase inhibitors block the action of an enzyme that allows HIV's RNA to be converted to DNA. (c) Integrase inhibitors prevent HIV's genetic material from plugging itself into the host cell's DNA. (d) Protease inhibitors prevent the breakdown of proteins necessary for assembly of new viral particles. *(Adapted from Health Resources and Services Administration HIV/AIDS Bureau, "HRSA Care Action," January 2007. Available online. URL: http://hab.hrsa.gov/publications/january2007/#a2. Accessed on July 15, 2008)*

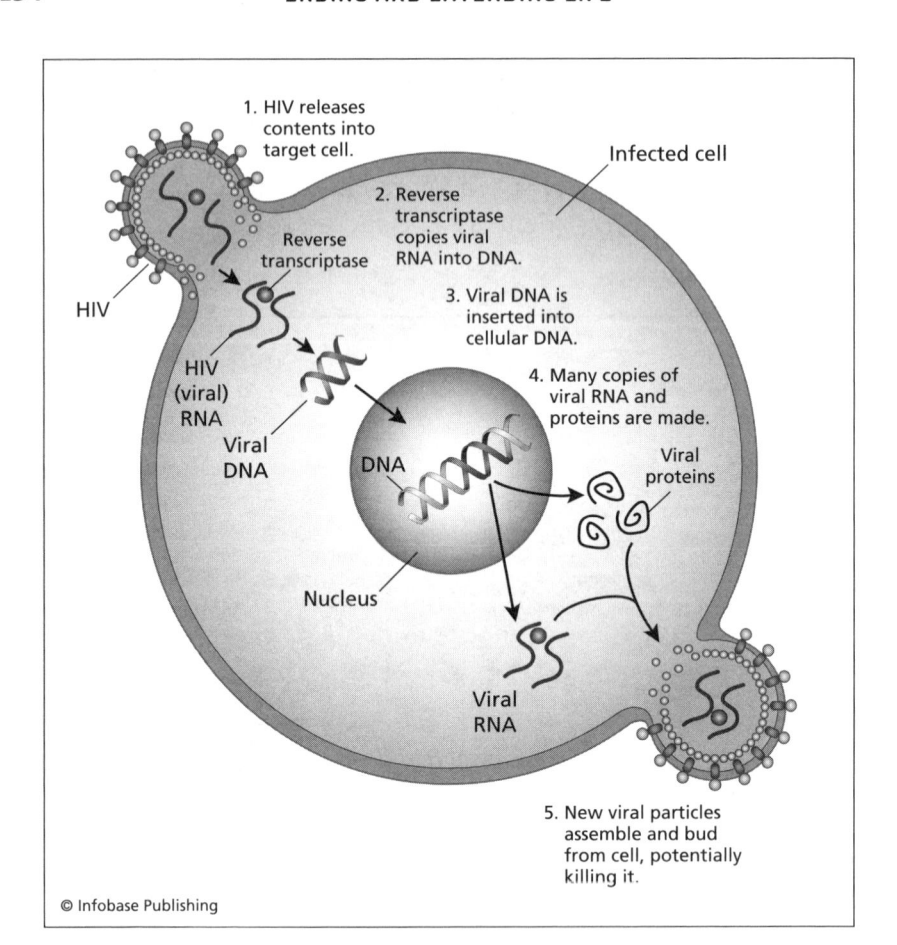

How HIV reproduces

Approaches to Prevention

In many countries, including the United States, social *stigma* and inadequate funding have hampered HIV prevention efforts. Some countries, however, have shown considerable success in slowing the spread of the disease through aggressive, science-based prevention programs. Uganda was a pioneer in the HIV/AIDS prevention effort with its "ABC" program (Abstinence, Be faithful, and use Condoms). In 1992, 16 percent of the population was infected; now 4 to 6 percent test HIV-positive.

There have been political impediments to Uganda's progress, however. During the past few years, the "C" (for condom) in

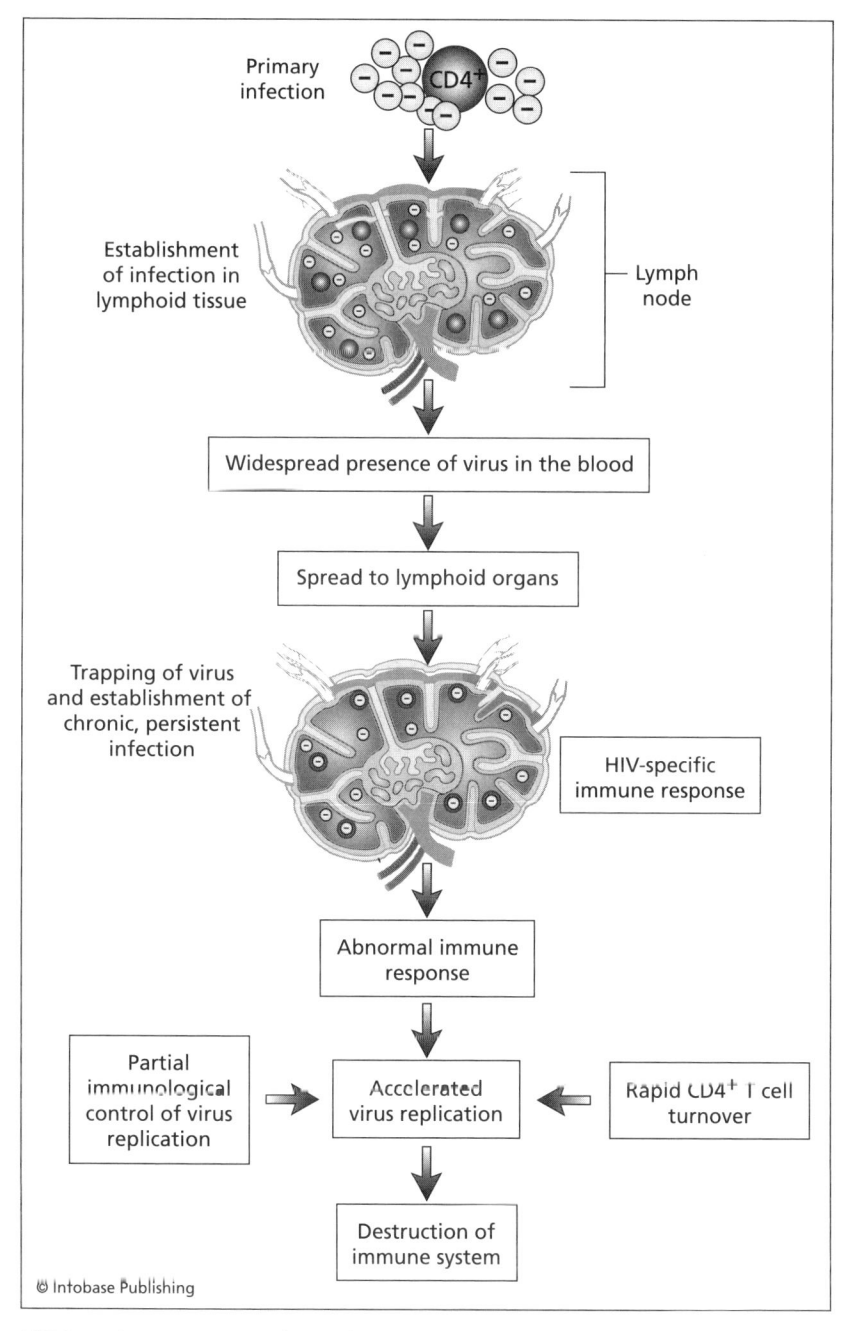

HIV pathogenesis—how the retrovirus attacks the body *(Adapted from A. Fauci, "HIV and AIDS: 20 Years of Science," Nature Medicine 9, no. 7 [2003]: 839–843)*

ABC has been downplayed and the "A" (for abstinence) has been emphasized, despite evidence that condom use has played a central role in the country's HIV reduction rate. The shift has been linked to funding from the United States (see sidebar, "U.S. Funding for HIV Treatment and Prevention" on page 148) and President George W. Bush's requirement that one-third of U.S. prevention money go to abstinence-until-marriage programs. In recent years, the rate of infection leveled off for Uganda's rapidly growing population, meaning that an increased number of people contract the HIV virus each year.

Thailand, with its 100 Percent Condom campaign, reduced new infections from about 140,000 in 1991 to 21,000 by 2003. Britain's needle-exchange programs have reduced the number of new infections from intravenous drug use to 6 percent. In the United States, by contrast, where federal funding for clean-needle programs has been banned, intravenous drug use accounts for about 22 percent of new infections.

Brazil's low rate of infection—0.6 percent—has been attributed to widespread HIV education and needle-exchange programs and has allowed that country to be the first in the world to provide free HIV drugs to its citizens who need them.

SUMMARY

Medical advances over the past 50 years have come at a high price, which a growing number of people find themselves unable to pay. A February 2007 poll by the *New York Times*/CBS News showed that a majority of Americans—nearly two-thirds—thought the federal government should guarantee health care for all Americans. When asked, "What if that meant that the cost of your own health insurance would go up?" 48 percent of those polled said they would still support universal health care. Sixty percent said that they would be willing to pay more taxes to ensure that everyone was covered, and half said they would pay up to $500 more in yearly taxes. Eight out of 10 Americans felt that guaranteeing health care was more important than extending recent tax cuts. An overwhelming majority supports the State Children's Health Insurance Program, with 84 percent of Americans saying they would expand the program to cover all children who are uninsured.

What about public support for U.S. funding of international health initiatives? On average, Americans believe that their country spends about 24 percent of the federal budget on assistance to developing nations. They also believe, on average, that this number should be more like 10 percent. In reality, the United States dedicates less than 1 percent of its total budget to aid for developing countries, including HIV/AIDS treatment and prevention.

According to the Global Forum for Health Research, of the more than $105.9 billion (in U.S. dollars) spent globally in 2001 on health research and development, an estimated 10 percent went to address the health problems that are plaguing 90 percent of the world's people. This disparity—known as the "10/90 gap"—has persisted at least since 1990, when it was first identified. And when it comes to drug development, the picture is bleaker. A 1999 study by Doctors Without Borders found that of the 1,393 new drugs approved between the years 1975 and 1990, a mere 13 of these—not quite 1 percent—were intended to treat infectious diseases that disproportionately affect low-income nations.

The impact of the 10/90 gap on health in low-income nations is staggering. Thanks largely to the availability of effective drugs and vaccines in wealthier countries, the death rate due to infectious diseases has plunged to one in 10. Among the world's poorest people, however, a staggering six in 10 deaths are still due to infectious diseases, many of which are preventable.

8

Health, Disease, and Wellness

How the medical establishment defines health and disease has major implications for patients and their families. People can mean different things by "healthy" and "sick," and where people's definitions are especially at odds, some members of a society may not receive needed medical care or may be unjustly forced to submit to treatment.

This chapter will look at cases—three past, three present—where mainstream cultural assumptions about what counts as healthy have determined what medical professionals and hospitals will treat, who they will treat, and how.

DEFINING HEALTH AND DISEASE

If a physician in 21st century America were to tell his young female patients to drop out of school because studying harmed their physical health, his behavior would be considered scandalous. He might be driven out of practice. Yet scarcely more than a century ago, physicians in mainstream American medical practice were doing just that.

A young girl's brain and her ovaries, doctors postulated, could not develop simultaneously. This speculation—which sounds outrageous today—was in line with the pragmatic concerns of a society utterly devastated by the Civil War, at a time when plummeting birthrates meant that female reproductive health was prized. Intellectual pursuits on the part of young women were discouraged—even considered to be a cause of disease.

In 1877, the Board of Regents of the University of Wisconsin put it this way: "[I]t is better that the future matrons of the state should be without a University training than that it should be produced at the fearful expense of ruined health; better that the future mothers of the state should be robust, hearty, healthy women, than that, by over study, they entail upon their descendents the germs of disease."

Elizabeth Blackwell, M.D. (1821–1910), the first woman to receive a medical degree in the United States *(AP Photo)*

Women in 21st-century America can be comforted that such archaic beliefs about the relationship between a woman's body and her mind no longer prevail in the mainstream of American medical practice, but it is notable that little more than a century ago, the idea that just thinking would harm a woman's ability to reproduce was part of the accepted body of scientific knowledge.

Similarly, just over a century ago, physicians in the mainstream of medical practice believed that African Americans were a biologically weaker race, physically and mentally inferior and more susceptible to disease than whites. Doctors, like many of their late-19th-century intellectual contemporaries, often were *Social Darwinists* who believed that stronger (Caucasian) races would prevail over weaker (African and Native American) races in the evolutionary struggle for survival.

"The weakest members of the social body are always the ones to become contaminated, and sooner or later succumb to the devitalizing forces of intemperance, disease, and crime and death," wrote physician J. Wellington Byers in a 1888 article, "Diseases of the Southern Negro" (published in *Medical and Surgical Reporter,* a professional journal of the time). Attitudes like these had moderated somewhat by the time the Tuskegee syphilis study was conceived, though the influences of Social Darwinism are clear in the way that the Tuskegee doctors talked about their patients. (See chapter 1 for an in-depth discussion of the infamous Tuskegee syphilis study.)

Perhaps the most dramatic shift in medical thinking about a minority group occurred in 1973, when the American Psychiatric Association radically altered its position on homosexuality by concluding that it was not, in fact, associated with any psychiatric or emotional problem whatsoever, and by removing it from the official manual of mental and emotional disorders. Before that time, it was standard practice for doctors to recommend psychotherapy to "cure" the "disease" of homosexuality. Two years later, the American Psychological Association endorsed the change.

"Psychologists, psychiatrists and other mental health professionals agree that homosexuality is not an illness, mental disorder or an emotional problem," says the current American Psychological Association statement on homosexuality. "Over 35 years of objective, well-designed scientific research has shown that homosexuality, in and of itself, is not associated with mental disorders or emotional or social problems." On the issue of choice, the association's position is that "human beings can not choose to be either gay or straight . . . [P]sychologists do not consider sexual orientation to be a conscious choice that can be voluntarily changed."

With the benefit of hindsight, it is clear that long-accepted medical opinions about women, African Americans, and homosexuals were shaped by cultural "norms" or standards, not by objective research. Any deviation from the average state of functioning (health) in a society is considered to be a disease only if it is seen by those in the mainstream as a "disvalue"—something to be avoided.

CONTEMPORARY DEBATES: WHAT IS FAIR, AND WHAT COUNTS AS TREATMENT?

Cultural norms are, by definition, much easier to identify looking back in time. The assumptions underlying current medical opinion are harder to discern because many people in mainstream society share those same beliefs.

The next two sections look at contemporary examples where cultural assumptions and stigmas have affected the lives of real patients. The first section considers reduced status for alcoholics on liver transplant lists, and the next looks at current efforts to equalize health care coverage for persons living with mental illness.

The final section of this chapter details a futuristic debate between mainstream physicians and academics on one side, and "transhumanist" thinkers on the other, about where legitimate health care needs stop and human "enhancement" begins. The medical possibilities often sound like the stuff of science fiction, but *human enhancement* is happening already. The question driving the debate is when, if ever, a society should deny its citizens the medical enhancements they feel will improve their lives.

Liver Transplants for Alcoholics

Alcoholic liver disease (ALD) is a catch-all category that includes *steatosis,* or "fatty" liver, which in itself is not harmful but can progress to more serious conditions; *alcoholic hepatitis,* a severe and sometimes fatal inflammation of the liver; and cirrhosis, a potentially deadly condition in which healthy liver cells are replaced with scar tissue. Cirrhosis is the fourth leading cause of death in Americans aged 45 to 54.

The liver is a large, complex organ and plays many vital roles in maintaining healthy body systems. It stores sugars, fats, vitamins, and other nutrients, and releases them as needed. It synthesizes bile, a substance necessary for digestion, as well as critical proteins including blood-clotting agents. It filters toxins, waste materials, and some infectious organisms from the blood. It is the principal organ in alcohol detoxification and is particularly susceptible to alcohol's harmful effects.

An estimated 10 to 20 percent of heavy drinkers develop alcoholic hepatitis, which may then progress to cirrhosis. But these deadly diseases are not limited to people with a chronic drinking problem. According to the Mayo Clinic, alcoholic hepatitis "can occur in people who drink only moderately or binge just once."

Alcohol-related end-stage liver disease (ARESLD) refers to toxic damage to the liver so severe that a liver transplant may be the only option. Due to a critical shortage of donated livers, a long-standing source of controversy is whether alcoholics should compete equally for liver transplantation. (Non–alcohol-related conditions that may require liver transplants include chronic hepatitis B and C, liver cancer, and autoimmune diseases of the liver.)

Two major arguments—one moral and one medical—have dominated the debate. The moral argument presumes that alcoholism is a character flaw, not a disease, and therefore that alcoholics are less deserving of transplants than other patients. Many medical ethicists dismiss this argument on the grounds that even if it could be proven that alcoholism and other addictions are moral failings and not diseases, clinicians should never be asked to perform impossibly complex moral calculations to determine who is most deserving of a new organ.

"Moral evaluation," concluded the Ethics and Social Impact Committee of the Transplant and Health Policy Center, at Ann Arbor, Michigan, "is wisely and rightly excluded from all deliberations of who should be treated and how. Indeed, we do exclude it. We do not seek to determine whether a particular transplant candidate is an abusive parent or a dutiful daughter; whether candidates cheat on their income taxes or their spouses; or whether potential recipients pay their parking tickets or routinely lie when they think it is in their best interests."

The second argument—the medical argument—hinges on whether a new organ might somehow be wasted if the recipient started drinking again. Two medical concerns—first, that alcoholic relapse might damage the new liver, and second, that drinking might interfere with patients' ability to take their scheduled medications that protect against rejection—have lead some medical ethicists to argue against equal consideration for alcoholics. Mark Siegler with the University of Chicago and Alvin H. Moss with West Virginia University concluded in a 1991 piece in the *Journal of the American Medical Association* that "since not all can live, priorities must be established and that patients with ARESLD should be given a lower priority for liver transplantation than others with ESLD."

Neither medical concern, however, is supported by research. According to the U.S. Department of Health and Human Services, "relapse rates in patients following transplant are lower than in patients undergoing alcoholism treatment, and serious relapses that adversely affect the transplanted liver or the patient are uncommon. In contrast, patients who receive a transplant because of an infection with hepatitis B or C viruses typically experience disease recurrence and are more likely to lose the transplanted liver because of recurrence of these infections." As for the concern about a recipient's ability to follow the antirejection medication regimen, "Liver rejection rates are similar for patients transplanted for ALD and those transplanted for other types of liver disease, indicating comparable rates of compliance with the antirejection medications."

Despite this persuasive body of evidence, the American Liver Foundation notes that some medical centers still will not perform liver transplants on alcoholics because they believe that a substantial percentage of these patients will return to drinking.

Equal Treatment for Mental Illness

Research has long shown that mental illnesses are physical illnesses where the primary organ affected is the brain, yet many health plans provide less coverage for mental illness than for other types of health problems. Efforts to rectify this stark inequality have been under way for decades, but progress toward full parity (equal treatment) for mental health has been slow, despite the fact that an estimated one in four American adults—about 58 million people—suffer from a diagnosable mental disorder in a given year.

A major step toward mental health parity finally came in October 2008, when Congress passed legislation requiring that group health insurance plans for 50 or more employees cover mental and physical health equally. Dr. Katherine Nordal, the American Psychological Association's executive director for professional practice, said that the bill's passage means that larger group plans "can no longer arbitrarily limit the number of hospital days or outpatient treatment sessions, or assign higher copayments or deductibles for those in need of psychological services." The new law takes effect January 1, 2010.

Mental conditions are the leading cause of disability for people between the ages of 15 and 44. These conditions include *mood disorders* like depression and *bipolar disorder, anxiety disorders* like *panic disorder* and *post-traumatic stress disorder (PTSD), schizophrenia,* eating disorders, *attention deficit/hyperactivity disorder (ADHD), autism spectrum disorders,* and Alzheimer's disease.

In any given year, 9.5 percent of the population lives with a *depressive disorder*, which, according to the National Institute of Mental Health (NIMH), "can destroy family life as well as the life of the ill person. But much of this suffering is unnecessary.... Unfortunately, many people do not recognize that depression is a treatable illness." Depression in older adults is especially undiagnosed, as it can present with symptoms frequently dismissed as a normal part of aging.

Bipolar disorder, also known as *manic-depressive illness,* is a type of depression characterized by unusual shifts in mood, energy, and ability to function. Approximately 5.7 million American adults have bipolar disorder. The mortality rate for people with the disorder is higher than the mortality rate for people with most types of heart disease and many types of cancer. An estimated two-thirds of people with the disorder have not been properly diagnosed or treated, yet if diagnosed, 80 to 90 percent of people can be treated effectively with medication and psychotherapy, allowing them to lead full and productive lives.

Anxiety disorders as a group are the most common mental illness in America. More than 18 percent—about 40 million American adults—are affected each year by these debilitating conditions. Children and adolescents can also develop anxiety disorders, and nearly three-quarters of those with an anxiety disorder have their first episode by age 21.5. Fortunately, there

Health, Disease, and Wellness

are effective treatments for these disorders if they are properly diagnosed.

More than 30,000 people die by suicide in the United States each year. The great majority (an estimated 95 percent) of suicide victims have an inadequately treated or untreated mental illness. Suicide is the third leading cause of death nationally for young people between the ages of 15 and 24.

According to the U.S. Department of Health and Human Services, approximately two-thirds of those with diagnosable mental illnesses do not seek treatment due to a lack of health insurance coverage, restrictions on their mental-health coverage under many insurance plans, or fear of being stigmatized for their illness. Stigma remains a pervasive and potentially lethal barrier to recovery, as it keeps people from seeking the help they

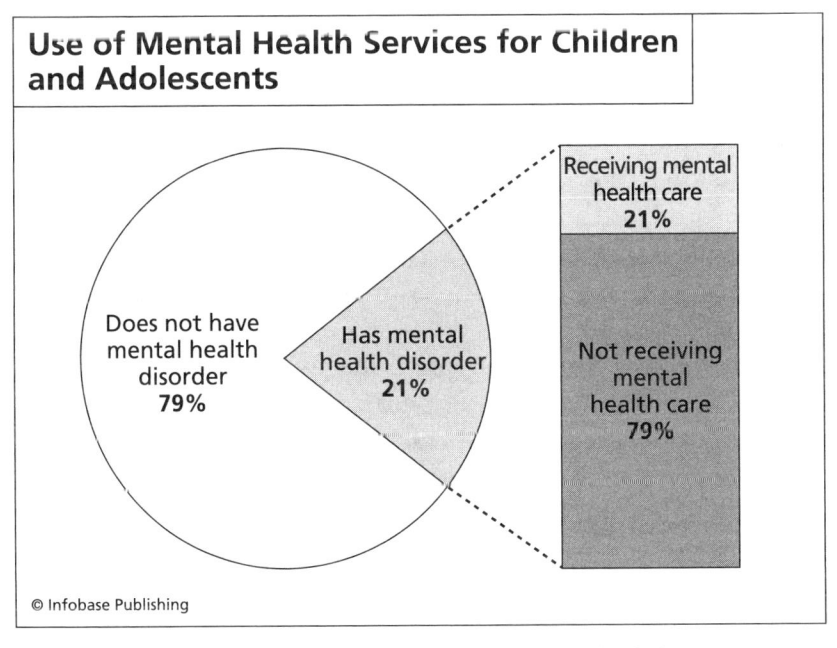

Use of Mental Health Services for Children and Adolescents

- Does not have mental health disorder: 79%
- Has mental health disorder: 21%
  - Receiving mental health care: 21%
  - Not receiving mental health care: 79%

© Infobase Publishing

Unmet mental health care needs of children and adolescents
(Adapted from Georgetown University's Center on an Aging Society, "Child and Adolescent Mental Health Services: Whose Responsibility Is It to Ensure Care?" October 2003. Available online. URL: http://hpi.georgetown.edu/agingsociety/pubhtml/mentalhealth/mentalhealth.html. Accessed July 15, 2008)

need due to fear of embarrassment or discrimination. A recent survey by the International Labour Organization (ILO) found that many American workers are "concerned about using their employee health benefits to obtain treatment for mental illness out of fear that their bosses and colleagues will learn about the problem and use it against them."

Until recently, stigma was the number one reason people did not seek treatment, but a new poll by the American Psychological Association showed that it was lack of insurance coverage that kept most respondents (87 percent) from seeking mental health treatment. Said Russ Newman, executive director for the American Psychological Association Practice Organization, "Health care coverage in this country needs to catch up with what people increasingly understand . . . [that] the mind and body are linked inextricably."

Unequal treatment of mental-health disorders in the United States dates back to the early 19th century, when the mental health treatment system was established as a separate entity from the mainstream health treatment system. In colonial times, people with mental illnesses were cared for by their families (if at all), but a population shift to the cities in the 19th century brought the problems of untreated mental illness into high relief. Isolated asylums were established, and the systematic separation of mental-health treatment from the rest of the health care system became the norm.

After World War II, hospitals began building psychiatric facilities, and health insurers began to include some inpatient mental-health care. Mental-health coverage gradually increased over the years, and now most employer-provided health care plans include some mental-health coverage. Yet coverage still is far from equal. According to the U.S. Department of Labor's Bureau of Labor Statistics, in the year 2000, 82 percent of employers with at least 100 workers still provided health insurance plans with more restrictive limits for mental health inpatient care (hospital room and board), and 90 percent imposed unequal limits on outpatient care.

One reason often cited for denying equal benefits for mental health is that it might drive up the cost of health care for employees and companies, but a recent study published in the *New England Journal of Medicine* suggests otherwise. The study looked at new mental health benefits with full parity for federal workers (ordered by President Bill Clinton in 1999, and available to all federal employees since 2001). The authors concluded that the expanded coverage "can improve insurance protection without increasing total costs" beyond plans that do not offer parity. One coauthor of the study, health economist Richard G. Frank of Harvard University, told the *New York Times* in March 2006 that "The big winners, in terms of reduced out-of-pocket spending, were the sickest patients, including those who needed hospital care."

Amanda L. Austin of the National Federation of Independent Business told the *Times* that she questioned the study's relevance to small businesses. "The Federal Employees Health Benefits Program, with a pool of nearly nine million members, is very different from the type of insurance plans available to a small business with three employees."

Senators Ted Kennedy and Pete Domenici, cosponsors of the Wellstone Mental Health Parity Act of 2003, introduce the bill in honor of the late Senator Paul Wellstone. *(Kristin Brooks Hope Center)*

Federal legislation known as the Paul Wellstone Mental Health Equitable Treatment Act (in honor of the late senator, who was a strong advocate for *mental health parity*) spent several years stalled in Congress, despite widespread bipartisan support. In early 2007, the issue received new life when both the House and Senate took up their own versions of parity legislation, with a compromise version finally becoming law as part of the massive economic rescue plan passed by both houses in October 2008. The bill extends equal coverage to all aspects of group health insurance plans for 50 or more employees, and it preserves state parity laws already on the books while extending coverage to 82 million Americans living in states without parity laws.

"Research shows," said Dr. Nordal, "that physical health is directly connected to emotional health and millions of Americans know that suffering from a mental health disorder can be as frightening and debilitating as any major physical health disorder. It's our hope that passage of this bill will force our health care system to finally start treating the whole person, both mind and body."

The Health versus Enhancement Debate

Health, as a concept, can be defined quite narrowly or quite broadly. *Merriam-Webster Medical Dictionary* defines it as "the condition of an organism or one of its parts in which it performs its vital functions normally or properly." The World Health Organization (WHO), on the other hand, defines the concept quite broadly as "a state of complete physical, mental, and social well-being and not merely the absence of disease or infirmity."

Pulitzer Prize–winning microbiologist René Dubos also defined health as something beyond the biological in his influential 1978 essay, "Health and Creative Adaptation," which appeared in the journal *Human Nature*. "For human beings," he wrote, "health transcends biological fitness. It is primarily a measure of each person's ability to do what he wants to do and become what he wants to become."

Critics of sweeping definitions like these say that they make it tough to distinguish between concerns about health and other desirable states like happiness. Ethicist Rosemarie Tong, in her book *New Perspectives in Healthcare Ethics,* argues that by the WHO standard, "most people probably are somewhat unhealthy." If health includes social well-being, then should cosmetic surgery procedures like nose jobs, liposuction, and hair or breast implants be considered standard treatment in societies focused on physical perfection? This question and others like it are being hotly debated in the medical ethics literature and mainstream media today, and the idea that physical and mental enhancement should be considered a basic health care right is gaining ground.

One southern California resident named Dennis Avner recently received media attention for spending over $150,000 on cosmetic procedures to look like a tiger. When health is measured by the ability to do what we want to do and become what we want to become (as proposed by Dubos), people draw different lines between what they think are legitimate health care needs and what they think are frivolous—or even harmful—medical procedures. Should Avner's surgically implanted whiskers or flattened forehead be considered standard care? Most mainstream medical ethicists say no, and categorize his medical choices as self-mutilation. Avner, on the other hand, does not believe that he is sick. Of Huron and Lakota descent, he says "in following a very old Huron tradition I am turning myself into my totem, a tiger."

An increasing number of people consider themselves part of a new movement called *transhumanism,* which seeks, in part, to challenge mainstream cultural assumptions about what should count as normal. Subscribers to transhumanism want to know why steroids for athletes are banned while muscle-enhancing surgery is permitted, and why children are allowed to drink caffeine but not to take other performance enhancing stimulants. "Transhumanists," according to the Transhumanist Declaration, "advocate the moral right for those who so wish to use technology to extend their mental and physical (including reproductive) capacities and to improve their control over their own lives." Drugs that enhance mental functioning, by this reasoning, should be available not only to people with an illness or disability (see sidebar below) but to anyone wishing to improve his mental performance.

The World Transhumanist Association describes itself as a nonprofit group that "advocates the ethical use of technology to expand human capacities." The organization supports "the development of and access to new technologies that enable everyone to enjoy *better minds, better bodies* and *better lives*. In other words, we want people to be *better than well*." Specifically, the organization supports the improvement of the human species (and in some cases, other species) through the use of "present

DRUGS TO TREAT CHILDHOOD BEHAVIOR

The hot-button topic of childhood behavior brings ideas about health and enhancement into high relief. When should a child's behavior be considered pathological, and when should it be altered with drugs? Some parents and experts believe that children who might be labeled "hyperactive" are just acting like normal children and should be left alone. Others think that when active children become disruptive, they should be treated with medication. Some transhumanist thinkers stretch the debate even further, seeing no reason why attention-enhancing drugs should not be made available to any person who wants them, whether or not that person is otherwise considered "normal."

Then there is aggression: When does hostile behavior become serious enough to be treated as a disorder? Some doctors prescribe drugs when a child intentionally hurts other children or their own parents emotionally or physically; others prescribe drugs when a child shows too much resistance to authority (a diagnosis known as *oppositional defiant disorder*). Shy behavior is another example: Shyness in children was technologies, such as genetic engineering, information technology, and pharmaceuticals, as well as anticipated future capabilities, such as nanotechnology, machine intelligence, uploading, and space colonization."

In its 2003 report, "Beyond Therapy: Biotechnology and the Pursuit of Happiness," President George W. Bush's bioethics council recommended against using technology to give people powers they do not have naturally. But human enhancement is already happening—some military pilots use the stimulant drug modafinil, for example—and an increasing number of mainstream medical ethicists want to sit down with transhumanists and talk about it.

In May 2006—the same weekend that the blockbuster supermutant movie *X Men: The Last Stand* hit movie screens—Stanford

---

once chalked up to temperament, but now it can be considered pathological and treated with personality-altering drugs to overcome social phobias.

The recent practice of treating difficult behavioral issues in children with "cocktails" of psychiatric medications has received much critical attention of late, particularly because the potential effects of drug mixes on children have not been studied in a systematic way. Dr. Thomas R. Insel, director of the National Institute of Mental Health, told the *New York Times* in November 2006, "There are not any good scientific data to support the widespread use of these medicines in children, particularly in young children where the scientific data are even more scarce."

But some psychiatrists swear by drug mixes and their potential benefits for young patients. Dr. Joseph Biederman, a professor of psychiatry at Harvard University, points out that doctors use drug mixes all the time to treat other medical conditions like heart disease and cancer. "Child psychiatry is not any different. These drugs have revolutionized how we treat severe psychopathology in children."

University's Center for Law and the Biosciences hosted a conference on human enhancement, and whether all humans should have a right to it. Walter Truett Anderson, then-president of the World Academy of Art and Science, delivered the keynote address and asked participants to look at the implications of human enhancement in a global context. Anderson and others believe that if human life spans are extended as transhumanists wish them to be, the natural environment might be taxed beyond repair. (See chapter 6 for ecological critiques of biogerontology—the scientific field that seeks to identify and delay the biological mechanisms of human aging.)

The time has arrived, says Anderson, for serious talk about the risks and benefits of technologically enhancing humans. "There are a lot of issues that are going to begin to surface," he told MSNBC. "People will have to confront them."

SUMMARY

Health coverage for the mentally ill, organ transplants for patients with addictions—these and other public health choices point to the importance of reevaluating cultural assumptions about health and disease in light of the latest scientific findings.

As an alternative to the terms "health" and "disease," some health care practitioners and ethicists choose to emphasize the concept of *wellness*—the quality of a person's life beyond a simple assessment of their state of health or disease. A person can be HIV positive, for example, and through a combination of medicines, alternative therapies, healthy lifestyle, and/or positive outlook feel quite well. A wellness approach tends to be more holistic in nature, emphasizing preventive care and treatment of the whole person. Wellness strategies are gaining in popularity, and they include things like meditation practice for the management of cancer pain, massage and yoga as routine preventive care for a variety of stress-related diseases, and healthy eating and exercise classes for people at risk for heart disease and diabetes.

Perhaps no other present-day debate challenges mainstream ideas about health and disease as provocatively as does the debate over human enhancement. Medical ethicist Eric Parens notes in his 2005 *Hastings Center Report* essay, "Authenticity and Ambivalence: Toward Understanding the Enhancement Debate," that as the discussion moves forward, many people will find that they feel morally ambivalent about some of the technological choices available to them. "If understanding is what we are after," he says, "we should embrace rather than suppress the ambivalence we often experience when we think about specific interventions."

Reflecting on that feeling of ambivalence—the place where moral certainty ends and doubt begins—is a solid strategy for exploring the ethical questions at the heart of all of the life-and-death issues considered in this volume.

CHRONOLOGY

MEDICAL RESEARCH ETHICS

1932 Tuskegee syphilis study begins

1945 In a secret army experiment, Ebb Cade is injected with 4.7 micrograms of plutonium during his stay at an Oak Ridge, Tennessee, hospital, likely without his knowledge or consent; thus begins a program of government-funded radiation experiments on humans

1947 An American military tribunal at Nuremberg, Germany, delivers its opinion on the ethical requirements for medical experimentation on human beings, subsequently known as the Nuremberg Code; this opinion codifies the notion of "voluntary consent" and forms the bedrock of future ethical codes for medical research with human subjects

1956 Study of hepatitis in children with mental retardation begins at the Willowbrook State School in Staten Island, New York

1958 Harry Harlow is elected president of the American Psychological Association

1963 The first animal-to-human transplants of the modern medical era, when Dr. Keith Reemtsma transplants chimpanzee kidneys into 13 patients; all patients die within nine months

1964 World Medical Association adopts the Declaration of Helsinki, an international code of ethics to guide doctors all over the world in research with human subjects

1966 Harvard anesthesiologist Henry K. Beecher publishes a landmark article in the *New England Journal of Medicine*

highlighting Willowbrook and 21 other ethically troubling experiments found in contemporary medical journals

Protections for animals are adopted by the federal government in the form of the Animal Welfare Act, a law protecting certain animals in some situations

1972 Story of the Tuskegee syphilis study breaks nationwide; the study is terminated

Willowbrook hepatitis study is terminated

1973 Senator Edward Kennedy and colleagues convene congressional hearings on human experimentation

1974 Federal regulations are promulgated requiring universities and other research institutions that seek or receive federal funds to maintain an Institutional Review Board (IRB), a body charged with reviewing the scientific and ethical merits of federally funded research

1975 Australian ethicist Peter Singer publishes his influential book, *Animal Liberation*, criticizing most animal experiments and highlighting Harry Harlow's experiments with primates

1979 The National Commission for the Protection of Human Subjects of Biomedical and Behavioral Research releases its landmark Belmont Report

1988 The U.S. Patent Office grants a patent to the President and Fellows of Harvard College for OncoMouse and other "transgenic non-human mammals"

1993 Kennedy Krieger Institute (KKI) begins a study of different methods of lead abatement and how well they reduce lead levels in the blood of inner-city children

1994 The U.S. Public Health Service recommends that the 076 antiretroviral regimen be given to HIV-positive pregnant women as standard treatment; controversial clinical trials of shorter-course therapies begin in low-income nations

1996 Federal regulations are revised to allow for some research with patients in emergency settings

1997 President Bill Clinton offers a formal apology to remaining survivors of the Tuskegee study

2000 The British animal-rights group Uncaged Campaigns publicizes experimental logs it has received that detail a five-year study of survival time of primates with genetically modified pig organs

2001 Maryland's highest court denounces the Kennedy Krieger Institute study of lead paint in children as unethical

2002 Senator Jesse Helms introduces an amendment to the Farm Bill excluding mice, rats, and birds—the majority of research animals—from the Animal Welfare Act

2003 Phase III clinical trial of the blood substitute PolyHeme begins

2006 Northfield Laboratories releases preliminary data on the PolyHeme study, with higher mortality in patients receiving the blood substitute than those receiving real blood; Northfield attributes the disappointing data to protocol violations

2008 The National Human Genome Research Institute, the National Toxicology Program, and the Environmental Protection Agency announce a joint effort to develop robotic molecular and cell-based approaches for determining the toxicity of thousands of chemicals; the plan is hailed by the Humane Society as a major step toward the replacement of animals in chemical testing

CRITICAL AND END-OF-LIFE CARE

1952 Bjørn Ibsen, a Danish anesthesiologist, saves a 12-year-old girl by introducing air into her lungs with a bag typically used to administer anesthesia; his method is adapted to design the modern mechanical ventilator

1957 Pope Pius XII declares that there is no obligation to use extraordinary means to prolong life and that it is up to doctors in individual cases to define the "moment of death"

1966 The term "locked-in syndrome" is introduced to describe paralysis of voluntary muscles brought on by damage to the brain stem

Chronology

1967 The first modern hospice, St. Christopher's, is founded in South London by Cicely Saunders

Christiaan Barnard performs the first human-to-human heart transplant in Cape Town, South Africa

1968 Harvard Medical School's Ad Hoc Committee to Examine the Definition of Brain Death releases its influential paper, "A Definition of Irreversible Coma," which defines irreversible coma as a new criterion for death

Juro Wada performs the first heart transplant in Japan; he is investigated for the murder of the organ donor and the recipient (who dies 83 days after surgery)

1972 The term "vegetative state" is introduced to describe patients who appear to be awake but who show no signs of awareness

Discovery of the powerful immunosuppressant drug cyclosporine makes organ transplantation a more viable means of care

1975 Karen Ann Quinlan falls into a coma at the age of 21 and is placed on a mechanical respirator; the New Jersey Supreme Court decides in favor of the parents' petition to have her removed from the machine, finding that they may assert a constitutional "right of privacy" on their child's behalf

1979 The first modern feeding tube, termed by one of its inventors the "percutaneous endoscopic gastrostomy," or PEG tube, saves the life of a 10-week-old infant unable to swallow

The Nobel Prize in medicine is awarded to Allan McLeod Cormack and Godfrey Newbold Hounsfield for the development of computerized axial tomography (CAT or CT scanning)

1983 Twenty-five-year-old Nancy Cruzan is diagnosed as persistently vegetative after her car flips on an icy, deserted road; a feeding tube is surgically implanted in her stomach

1985 Karen Ann Quinlan dies of pneumonia at the age of 31

1987 Nancy Cruzan's parents ask the hospital to remove her feeding tube, but the hospital requires a court order

1988	The Missouri Supreme Court denies Nancy Cruzan's parents' petition to withdraw her feeding tube
1990	In a landmark decision, the U.S. Supreme Court rules in favor of Missouri's (or any state's) right to require its own standards of evidence, while also upholding an individual's right to reject "lifesaving hydration and nutrition"
	Nancy Cruzan's feeding tube is removed legally later that year, nearly eight years after her car crash; she dies 12 days later
	Theresa Schiavo enters cardiac and respiratory arrest at the age of 26
1991	Fourteen-month-old Marissa Eve Ayala provides a lifesaving bone marrow transplant for her sister, Anissa
1994	Oregon passes a law, known as the Death with Dignity Act, which makes it legal for a physician to write a prescription for a lethal dose of medication for a terminally ill patient
1996	The "triple cocktail" treatment for HIV/AIDS is introduced
2001	Theresa Schiavo's feeding tube is removed the first time, upon order of a Florida judge; a different judge orders it reinserted two days later
2002	The term *minimally conscious* is introduced to describe people who formerly would have been described as vegetative but who can track movement with their eyes and seem intermittently responsive
2003	Terry Wallis, an Arkansas mechanic, recovers awareness, more than 18 years after a serious car accident
	Theresa Schiavo's feeding tube is removed a second time, but the Florida State Legislature passes "Terri's Law" and the tube is reinserted six days later
	The Nobel Prize in medicine is awarded to Sir Peter Mansfield and Paul Lauterbur for their discoveries in magnetic resonance imaging (MRI or MR scanning)
2005	Theresa Schiavo's feeding tube is removed a third time; she dies at the age of 41

Baylor Regional Medical Center in Plano, Texas, removes a terminally ill woman, Tirhas Habtegiris, from the mechanical ventilator against the wishes of her family, under authority of a controversial Texas law known as the Advance Directives Act

2006 An fMRI scanner measures unexpected mental activity in a 23-year-old British woman diagnosed as vegetative after a devastating car accident

GLOSSARY

active control trial a **clinical trial** in which a control group receives the standard treatment for a condition (rather than a placebo), and a treatment group receives the experimental treatment

active infection an infection in which the disease-causing agent is reproducing rapidly and spreading to other cells in the body

advance directive see **living will**

AIDS (acquired immunodeficiency syndrome) a collection of symptoms and infections resulting from damage done to the immune system by the human immunodeficiency virus (**HIV**)

alcoholic hepatitis a severe, sometimes fatal, inflammation of the liver due to alcohol use

alcoholic liver disease (ALD) a catchall category that includes alcohol-related steatosis ("fatty" liver), hepatitis, and cirrhosis

alcohol-related end-stage liver disease (ARESLD) alcohol damage to the liver so severe that a liver transplant may be the only option

Alzheimer's disease a degenerative brain disease characterized by loss of memory, impaired cognitive processes, and personality changes

anxiety disorder an umbrella term that includes a wide variety of medical disorders characterized by severe fear and anxiety; anxiety disorders affect an estimated 40 million American adults

artificial respirator see **mechanical ventilator**

attachment theory a theory of psychological development emphasizing the importance of infants' or children's relationships to their parents or other caregiving adults

Glossary

attention deficit/hyperactivity disorder (ADHD)   a condition that can become apparent in early life; a child with ADHD may have difficulty focusing or controlling impulses

autism spectrum disorders   a collection of medical conditions all characterized by mild to severe impairment in social interaction and communication and by repetitive behaviors or restricted patterns of interest

autonomy   the ability of a person to make independent choices

AZT   see ZDV

beneficence   one of three central **ethical principles** articulated in the Belmont Report; in medical ethics, the term refers to the intention on the part of medical providers to act in the best interests of their patients

best interests standard   a standard proposed by some medical ethicists for when to withdraw care from a vegetative or terminal patient who is no longer competent; the standard would allow for withdrawal of care regardless of the wishes of family members, provided that it is in the best interests of the patient

biogerontology   the study of the biological processes of aging

bipolar disorder   a medical condition that can cause severe shifts in a person's mood, energy, and ability to function

black fever   see **VL (visceral leishmaniasis)**

bone marrow   the soft tissue at the center of most bones that produces red and white blood cells and platelets

brain death   total and permanent loss of all brain function; one of the medical and legal definitions of death (along with heart-lung death)

bulimia   an eating disorder characterized by bouts of excessive eating followed by purging (through misuse of laxatives or self-induced vomiting), fasting, and/or excessive exercise

cancer   a category of diseases in which abnormal cells divide without control

CD4 cell   the chief target of the HIV virus, this specialized white blood cell regulates much of the immune response to infection; also called a helper T-cell

centenarian a person who has lived to the age of 100 or beyond

chemotherapy treatment with drugs to kill **cancer** cells and shrink tumors

chimera an organism composed of two genetically different cell or tissue types, named for the mythological creature that was part lion, part goat, and part serpent

chronic disease a long-lasting disease or condition; from the Greek *chronos,* meaning time

cirrhosis a potentially deadly condition in which healthy liver cells are replaced with scar tissue

clinical involving the direct observation and treatment of patients

clinical depression (major depressive disorder) a state of intense sadness or despair that has progressed to the point of requiring professional attention

clinical trial an experiment designed to test the effectiveness and safety of a new drug or treatment in humans

coma a state of profound unconsciousness—usually the result of injury, disease, or poison—in which a patient is incapable of waking or responding to external stimuli and internal needs (from the Greek *kōma,* meaning deep sleep)

congenital a condition present at birth; can be due to environmental or genetic factors

control group a group of participants in a clinical trial who receive either the standard treatment for a condition or a "dummy" treatment (**placebo**) instead of the experimental treatment

coronary artery disease the most common form of heart disease, in which sticky deposits (plaque) block adequate blood flow to the heart

critical function a biological system without which the whole organism cannot operate; e.g., breathing or circulation

CT (computerized axial tomography) **scan** diagnostic technology that uses computer algorithms to compile a series of two-dimensional X-ray images of the body into three-dimensional images

cyclosporine a strong drug that suppresses the action of the immune system, administered to prevent rejection of donated organs and to treat certain autoimmune diseases

DAF2 (decay accelerating factor) **gene** a gene believed to be important to the aging process

depressive disorder a broad term covering many medical conditions characterized by a sense of sadness, hopelessness, or worthlessness, and by a loss of interest in favorite activities; an estimated 20 million American adults live with depressive disorders, and examples include major depressive disorder (**clinical depression**), postpartum depression, and seasonal affective disorder (SAD)

diabetes a chronic disease in which the body is unable to properly regulate levels of sugar in the blood

DNA (deoxyribonucleic acid) the chain of molecules inside a cell that carries genetic information and passes it from one generation to the next; it is the chemical blueprint for building, running, and maintaining living organisms

DOT (diffuse optical tomography) imaging technology that uses light at the near-infrared part of the spectrum (light of a longer wavelength than visible light) to image different levels of oxygen in hemoglobin, the oxygen-carrying protein in blood

eating disorder any medical condition involving severe disturbances in eating behavior, such as extreme reduction of food intake or extreme overeating, or involving feelings of extreme distress or worry about body weight or shape

electrocardiogram (EKG or ECG) a test that records the electrical activity of the heart

electroencephalogram (EEG) a test that records the electrical activity of the brain as detected by electrodes placed on the scalp

embryonic stem cells cells from the early embryo with the potential to develop into the specialized cell types of the body

endemic disease a disease that is always present in a given population or area

endogenous retroviruses retroviruses thought to be incorporated in the DNA of humans and other vertebrates by way of ancient infection of germ cells (eggs and sperm)

endothelial cells the thin layer of cells that line the interior surface of blood vessels

endotracheal tube a tube that goes through the mouth into the windpipe to support breathing, usually in concert with mechanical ventilation

entry inhibitor a class of HIV drugs that work by preventing the virus from entering healthy **CD4 cells**; this drug category includes **fusion inhibitors**

enzyme a protein in the body that can speed up important chemical reactions

ER (estrogen receptor)**-positive** a tumor in which individual cancer cells have estrogen receptors on their surfaces and may use the hormone to reproduce

ethical principles general guidelines for ethically appropriate behavior; for example, the Belmont Report outlined three principles by which to evaluate medical research with human subjects (**respect for persons, beneficence,** and **justice**)

extubate remove a tube inserted in the airway (endotracheal tube)

feeding tube a tube used to deliver nutrients to patients who are unable to swallow; examples include the **PEG tube, NG tube,** and **jejunostomy tube**

FK-506 a strong drug that suppresses the immune system and is used to prevent organ transplant rejection; discovered in 1984

fMRI (functional MRI) a powerful new brain-mapping technology that measures blood-flow changes—and thus neurological activity—by measuring differences in blood-oxygen levels

fusion inhibitor a class of HIV drugs that block the virus from binding to receptors on a cell's surface, thus preventing HIV from entering and infecting the cell

futile treatment life-extending treatment that is deemed inappropriate because it will prolong suffering without meaningful hope of recovery, or—in the case of certain profound brain injuries—will sustain certain systems of the body without meaningful hope of regaining awareness

GAL (galactose-alpha,1,3-galactose) a sugar that exists on the surfaces of many common viruses and bacteria and that the human immune system is programmed to recognize and destroy; cells that line the insides of blood vessels in pig organs are also studded with this molecule

gamma globulin a protein component in the blood that contains antibodies

GDP (gross domestic product) the value of all goods and services produced by a country in a given year

genetic engineering artificially introducing changes to an organism's or cell's genetic material

healthy volunteer a medical research study participant who is not a patient

heart disease a general term for a number of conditions that affect the heart—the most common of them coronary artery disease, which is the leading cause of death in the United States

heart-lung death death declared on the basis of prolonged absence of heartbeat and breathing, rather than on the basis of neurological criteria

helper T-cell see **CD4 cell**

hemorrhagic shock shock due to severe blood loss

hepatitis inflammation of the liver, caused by infection or toxic agents

HIV (human immunodeficiency virus) a retrovirus that attacks and damages the immune system, the cause acquired immune deficiency syndrome (AIDS); it is transmittable through sexual intercourse or contact with infected blood

HIV-1 M the most common and virulent form of HIV, traced to chimpanzee populations in southern Cameroon

HIV-2 a less virulent strain of HIV occurring largely in West Africa and linked to a monkey called the sooty mangabey

hospice a special program of care and emotional support for people in the final phase of illness, as well as for their families and caregivers; hospice care may take place in the patient's home or in a hospital setting

host cell a cell that is infected by a virus or other microorganism and provides metabolic support to the foreign agent

human enhancement a disputed term that usually refers to any attempt to overcome current limitations of the human body; medical ethicists typically restrict the use of this term to any technological intervention in human biology that is not intended to treat an illness or disability

Huntington's disease a genetic, degenerative disease of the brain and central nervous system that causes progressive loss of mental function and motor control

hyperacute rejection an immediate and violent immune reaction that quickly kills a newly transplanted organ

immune response the activity of the immune system against foreign substances or organisms in the body

immunosuppressive drug (immunosuppressant) a drug used to suppress the immune system to prevent rejection of transplanted organs or tissues, or to treat certain autoimmune diseases

informed consent voluntary consent by a patient or healthy volunteer to participate in a medical research trial, or voluntary consent by a patient to undergo medical treatment, after having been fully informed of—and (ideally) having understood—the anticipated risks and benefits of the treatment or experiment

integrase inhibitor a class of HIV drugs that prevent HIV from inserting its genetic material into a host cell's normal DNA

intracranial hemorrhage bleeding within the skull

ionizing radiation high-energy radiation (e.g., ultraviolet rays, X-rays, and gamma rays) that can remove electrons from atoms and create charged (excited) particles, potentially damaging the molecular structure of cells

IRB (institutional review board) a committee formally charged with ethical and scientific review of biomedical and behavioral research involving human subjects; IRB review is required for all human-subject research receiving federal funds

irreversible coma a coma (profound unconsciousness) due to brain damage so severe that a patient is not expected to regain consciousness; this is a diagnosis that is typically made based on results of standard bedside tests

jejunostomy tube a feeding tube surgically placed in the small intestine

justice one of three central ethical principles articulated in the Belmont report; in medical ethics, the term usually refers to distributive justice—the idea that social benefits and burdens should be shared fairly

lead abatement the reduction or elimination of hazards due to lead contamination

lead poisoning a toxic condition due to the absorption of excessive levels of lead

leukemia cancer of the body's blood-forming tissues, in which the **bone marrow** produces a large amount of abnormal white blood cells that do not function properly; *leukemia* means "white blood" in Greek

living will a legal document expressing a person's decisions about the use of artificial life support in the event that she is unable to speak for herself; also called an advance directive

locked-in syndrome a condition caused by damage to the brain stem, in which a patient is aware but cannot move due to paralysis of voluntary muscles

malaria a disease caused by a parasite that infects red blood cells and usually is transmitted to humans by mosquitoes

manic-depressive illness see **bipolar disorder**

mechanical ventilator a machine that assists or maintains breathing for patients unable to breathe on their own; also called an artificial respirator

medical proxy someone appointed by a patient to make medical decisions if the patient is incapacitated

mental health parity in health care policy, this is the idea that there should be equality of medical coverage for mental and physical illnesses

metastatic cancer cancer that has spread from its primary site to other places in the body

microbicide any substance that destroys or disables infectious microbes

minimally conscious a term used to describe some people who formerly would have been described as vegetative, but who can track movement with their eyes and seem intermittently responsive

minimal risk when the degree of risk from taking part in a medical experiment is thought to be small—no greater than the risks typically encountered in daily life or in the course of routine medical examinations or tests

mood disorder a group of medical disorders characterized by disturbance or instability of mood, and including **depressive disorders** and **bipolar disorder**

MRI (magnetic resonance imaging) imaging technology that uses a powerful magnetic field to align excited water molecules with or against the field, and then analyzes the pattern of absorption and transmission of radio waves by those water molecules to create images of the body

mutation any change in the DNA of a cell (harmful, beneficial, or with no effect); mutations may be caused by mistakes during cell division or by exposure to DNA-damaging agents in the environment

myelogenous leukemia a disease in which an abnormal protein causes the bone marrow to produce too many white blood cells (all of which contain the mutated chromosome that produce the abnormal protein); eventually the abnormal cells can crowd out healthy blood cells

nanotechnology a field incorporating scientific advances in a wide range of topics dealing with matter on a very small scale, including protein synthesis, molecular engineering, and micro-computing

neuroimaging imaging studies of the brain and nervous system

NG (nasogastric) **tube** a feeding tube inserted into the nose, through the throat and esophagus and into the stomach

nontherapeutic experiment an experiment that is not expected to bring therapeutic benefit to participants

oncogene a modified gene capable of transforming normal cells into cancer cells

oppositional defiant disorder a controversial pediatric diagnosis characterized by an ongoing pattern of tantrums, arguing, and hostile/disruptive behaviors toward parents, teachers, and other authority figures

palliative care treatment to improve a patient's quality of life by relieving—rather than attempting to cure—symptoms of a chronic or terminal illness

pandemic an outbreak of disease that spreads over a large region, even throughout the world

panic disorder a medical condition characterized by sudden, intense panic attacks

Pap test a method of examining cells of the cervix to detect cancerous and precancerous cells; also called the *Pap smear*

PEG (percutaneous endoscopic gastrostomy) **tube** a feeding tube surgically implanted in the stomach through the abdominal wall

penicillin an antibiotic produced by mold and used to rid the body of many kinds of bacterial infections

perinatal the period just before, during, and after birth

persistent vegetative state (PVS) a condition resulting from severe damage to the thinking, feeling part of the brain (the cerebral cortex), in which a patient may demonstrate sleep-wake cycles without detectable awareness

PERV (porcine endogenous retroviruses) ancient **retroviruses** long ago incorporated into the pig's genetic code, and a source of concern in cross-species transplantation because of the chance that they might become reactivated in humans

PET (positron emission tomography) **scan** imaging technology that takes biologically active chemicals (such as glucose to detect cancerous tumors, or neurotransmitters like serotonin to study neurological illnesses) and tags them with a radioisotope to measure metabolic (uptake) processes in the body

physician-assisted suicide (PAS) when a physician assists a patient in ending his or her own life; physicians in Oregon and Washington may perform PAS legally by writing a prescription for a lethal overdose of drugs

placebo an inactive substance or dummy treatment administered to one group of experimental subjects to provide a comparison to the real effects of a test drug or treatment administered to a different group

post-traumatic stress disorder (PTSD) an anxiety disorder that can develop after exposure to a terrifying event, and may involve flashbacks and a range of intense physical and psychological symptoms

prolongevity extension of the human life span

protease inhibitor a class of HIV drugs designed to block the action of the enzyme HIV protease, which aids in HIV replication

protocol a written plan for a clinical trial or other experiment, which states the purpose of the experiment and exactly how the study will be conducted

radiation therapy the use of high-energy radiation (e.g., X-rays, gamma rays, neutrons, or protons) to kill cancer cells and shrink tumors

respect for persons one of the ethical principles articulated in the Belmont Report, which includes respect for the autonomy of people who are capable of making informed, independent decisions and special protections for people whose autonomy is impaired (either due to incompetence or the potential for coercion)

retrovirus a type of virus that contains **RNA** instead of **DNA** and is able to incorporate itself into the DNA of the host cell

reverse transcriptase inhibitor a class of HIV drugs that work to block the action of the enzyme reverse transcriptase, which allows HIV's **RNA** to be converted to **DNA**

RNA (ribonucleic acid) one of the two types of nucleic acids (**DNA** is the other) found in all cells; RNA contains genetic information necessary to synthesize specific proteins

salvarsan (arsphenamine) the first modern chemotherapeutic agent; it was used to treat **syphilis** in the early 20th century, often with toxic side effects

schizophrenia a chronic, severe brain disorder in which affected people may hear voices that other people do not hear or may believe that others are reading their minds, controlling their thoughts, or plotting against them

sentience the possession of sense organs and a nervous system capable of feeling or perceiving sensory input

Social Darwinism the improper application of the biological concept of natural selection to the historical development of human societies; the idea that some races or types of people are more fit to survive—and to thrive—than others

speciesism a term used to describe prejudice against other animals, similar to racism or sexism

SPECT (single photon emission computed tomography) imaging technology that employs gamma-ray emitting radioisotopes; it is particularly useful when an especially rapid "snapshot" of blood flow to the brain is needed

spinal tap a procedure in which a fine needle is inserted between two vertebrae in the lower part of the spine to collect cerebrospinal fluid or to administer drugs; also called a *lumbar puncture*

spirochete a bacterium with a long, thin, coiled (spiral) shape

standard of care treatment or care that is considered by medical experts to be appropriate, commonly accepted, and widely used

STD (sexually transmitted disease) infections that are most commonly spread through sexual contact; also called *venereal disease*

steatosis also known as "fatty" liver, this condition is not thought to be harmful in itself but may progress to more serious conditions

stem cell an unspecialized cell that can, under the right conditions, develop into the specialized cells of the body

stigma a negative stereotype about a group of people; of special concern in medical care when fear of stigma prevents people with certain conditions from seeking medical treatment (e.g., people with mental illnesses or sexually transmitted diseases)

synthetic blood substitute a synthetic substance that can carry oxygen and that may serve as a temporary replacement for blood

syphilis a sexually transmitted disease caused by the *T. pallidum spirochete* (bacterium) and treatable with penicillin; if left untreated, syphilis can cause serious and sometimes fatal damage to the central nervous and cardiovascular systems

terminal illness a disease diagnosed as incurable and expected to result in death

terminal sedation (TS) the treatment of pain in terminally ill patients, even to the point of causing unconsciousness or hastening death, usually by means of continuous intravenous administration of a sedative drug; also known as *palliative sedation*

therapeutic any procedure or drug that is expected to provide medical benefit to patients or alleviate their symptoms

tracheostomy tube a tube inserted in an opening in the trachea to assist with breathing; commonly called a "trach"

transgenic animal an animal whose genetic makeup has been altered by the introduction of genetic material from another species

transhumanism a movement that advocates the physical and mental enhancement of humans (and of some other animals) by any available technological means

triple cocktail an extraordinarily effective treatment for HIV/AIDS that combines a protease inhibitor with two reverse transcriptase inhibitors; the triple cocktail treatment was introduced in 1996

universal health care a national health care system that ensures all citizens free or subsidized access to health care services

utilitarianism the ethical theory that seeks the greatest benefit for the greatest number

ventricular tachycardia an abnormally rapid heart rhythm that originates in the lower chambers of the heart (ventricles)

vertical chamber apparatus a device used in experiments on rhesus macaque monkeys at the University of Wisconsin during the 1970s to produce an animal model of clinical depression

vertical transmission transmission of an infectious disease from a mother to a fetus or baby during pregnancy, birth, or breast-feeding

viral encephalitis inflammation of the brain resulting from a viral infection

VL (visceral leishmaniasis) a deadly parasitic disease, transmitted by the bite of an infected sand fly; also known as *kala-azar* or *black fever*

vulnerable population a group of research subjects who may require special protections to ensure that they are treated ethically

waived-consent trial a special exception to informed consent rules for certain studies in emergency medicine, in which patients unable to consent for themselves can be used in research under certain carefully controlled conditions; requirements for such studies include the prospect of direct benefit for participants in a life-threatening situation and the lack of availability of other proven treatments

wellness in alternative medicine, this term refers to an overall quality of physical and mental life beyond a simple assessment of health and disease; a goal that encourages treatment of the whole person, not just symptoms

xenotransplantation transplantation of organs or tissues from an organism of one species into an organism of a different species

ZDV (zidovudine) the first antiretroviral drug approved for the treatment HIV, now often used in combination with other drugs; also called AZT

FURTHER RESOURCES

Chapter 1: Ethical Principles in Medical Research
Information for this chapter is drawn largely from press reports and journal articles about the Tuskegee syphilis study and the HIV vertical transmission trials; historical documents such as the Nuremberg Code, the Declaration of Helsinki, and the Belmont Report; and James H. Jones's in-depth treatment of Tuskegee, *Bad Blood: The Tuskegee Syphilis Experiment*. All sources are detailed below.

Brandt, Allan M. "Racism and Research: The Case of the Tuskegee Syphilis Study." *Hastings Center Report* 8, no. 6 (December 1978): 21–29.
French, Howard W. "AIDS Research in Africa: Juggling Risks and Hopes." *New York Times* (October 9, 1997). Available online. URL: query.nytimes.com/gst/fullpage.html?res=9D04E0D9143CF93AA35753C1A961958260. Accessed December 7, 2007.
John F. Kennedy School of Government, Harvard University. "The Debate over Clinical Trials of AZT to Prevent Mother-to-Infant Transmission of HIV in Developing Nations." Available online. URL: www.ksg.harvard.edu/case/azt/ethics/home.html. Accessed December 7, 2007.
Jones, James H. *Bad Blood: The Tuskegee Syphilis Experiment*. New York: The Free Press, 1993.
Mitchell, Alison. "Clinton Regrets 'Clearly Racist' U.S. Study." *New York Times* (May 17, 1997). Available online. URL: query.nytimes.com/gst/fullpage.html?res=9504E1D81338F934A25756C0A961958260. Accessed December 7, 2007.
National Commission for the Protection of Human Subjects of Biomedical and Behavioral Research. "The Belmont Report: Ethical Principles and Guidelines for the Protection of Human Subjects of Research," 1979. Available online. URL: ohsr.od.nih.gov/guidelines/belmont.html or www.hhs.gov/ohrp/belmontArchive.html. Accessed April 30, 2008.

National Institute of Allergy and Infectious Diseases, National Institutes of Health. "Preventing HIV Infection in Infants in Developing Countries: NIAID's Role." Available online. URL: www.niaid.nih.gov/publications/agenda/1197/page2.htm. Accessed December 7, 2007.

Nuremberg Code. From "Permissible Medical Experiments," *Trials of War Criminals before the Nuernberg Military Tribunals under Control Council Law No. 10: Nuernberg, October 1946–April 1949,* vol. 2: 181–184. Washington, D.C.: U.S. Government Printing Office, n.d. Available online. URL: www.loc.gov/rr/frd/Military_Law/pdf/NT_war-criminals_Vol-II.pdf. Accessed December 7, 2007.

Office of Human Subjects Research, National Institutes of Health. "IRB Protocol Review Standards." *Guidelines for the Conduct of Research Involving Human Subjects at the National Institutes of Health.* April 2004. Available online. URL: ohsr.od.nih.gov/guidelines/GrayBooklet82404.pdf. Accessed November 4, 2008.

Stolberg, Sheryl Gay. "U.S. AIDS Research Abroad Sets Off Outcry over Ethics." *New York Times* (September 18, 1997). Available online. URL: query.nytimes.com/gst/fullpage.html?res=9D04E7DB1038F93BA2575AC0A961958260. Accessed December 7, 2007.

Wooten, James T. "Survivor of '32 Syphilis Study Recalls a Diagnosis." *New York Times* (July 27, 1972).

World Medical Association. "Declaration of Helsinki," 1964. Available online. URL: history.nih.gov/laws/pdf/helsinki.pdf. Accessed December 7, 2007.

World Medical Association. "Declaration of Helsinki," amended. Available online. URL: www.wma.net/e/policy/pdf/17c.pdf. Accessed December 7, 2007.

Chapter 2: Vulnerable Populations

For issues arising with medical research on vulnerable populations, Ronald Munson's *Intervention and Reflection: Basic Issues in Medical Ethics* and Bonnie Steinbock et al.'s *Ethical Issues in Modern Medicine* are key sources for this chapter. Historical sources include Henry Beecher's seminal 1966 article, "Ethics and Clinical Research," as well as government documents uncovered by the President's Advisory Committee on Human Radiation Experiments. All sources are detailed below.

Advisory Committee on Human Radiation Experiments. *The Human Radiation Experiments: Final Report of the President's Advisory Committee.* New York: Oxford University Press, 1996.

Altman, Lawrence K. "Immunization Is Reported In Serum Hepatitis Tests." *New York Times* (March 24, 1971).

Beecher, Henry K. "Ethics and Clinical Research." *The New England Journal of Medicine* 274, no. 24 (June 16, 1966): 1,354–1,360. Available online. URL: www.who.int/bulletin/archives/79(4)367.pdf. Accessed December 7, 2007.

Bridges, Andrew. "Experts: Tell Public about Trauma Tests." Associated Press (October 11, 2006).

Buchanan, David R., and Franklin G. Miller. "Justice and Fairness in the Kennedy Krieger Institute Lead Paint Study: The Ethics of Public Health Research on Less Expensive, Less Effective Interventions." *American Journal of Public Health* 96, no. 5 (May 2006): 781–787.

Burton, Thomas M. "Red Flags: Amid Alarm Bells, a Blood Substitute Keeps Pumping." *Wall Street Journal* (February 22, 2006).

———. "Blood-Substitute Study Is Criticized by U.S. Agency." *Wall Street Journal* (March 10, 2006).

———. "Grassley Accuses FDA of Laxity in Study of Blood Substitute." *Wall Street Journal* (March 14, 2006).

———. "SEC Begins Informal Probe of Northfield Labs Over Blood Studies." *Wall Street Journal* (March 17, 2006).

———. "Use of Substitution for Blood Draws Ethics Challenge." *Wall Street Journal* (March 20, 2006).

———. "Northfield Hits Blood-Test Hurdle; 'Preliminary' Findings of PolyHeme Death Rate Suggest Approval Setback." *Wall Street Journal* (December 20, 2006).

———. "FDA Nod Sought for Blood Substitute." *Wall Street Journal* (August 15, 2007).

Cohn, Victor. "Vaccine Is Developed for Serum Hepatitis." *Washington Post* (March 24, 1971).

Duke University Medical Center News Office. "Blood Substitute to Be Tested at Duke University Hospital." May 13, 2004. Available online. URL: www.dukemednews.org/news/article.php?id=7607. Accessed December 9, 2007.

Feder, Barnaby J. "Mixed Test Results for Blood Substitute; Maker's Shares Fall." *New York Times* (December 20, 2006). Available online. URL: www.nytimes.com/2006/12/20/health/20blood.

Further Resources

html?_r=1&adxnnl=1&oref=slogin. Accessed February 25, 2008.

Hornblum, Allen M. *Acres of Skin: Human Experiments at Holmesburg Prison.* New York: Routledge, 1998.

Johns Hopkins Medicine, Office of Communications and Public Affairs. "Kennedy-Krieger Institute Lead-Based Paint Study Fact Sheets." September 7, 2001. Available online. URL: www.hopkinsmedicine.org/press/2001/SEPTEMBER/010907.htm. Accessed December 9, 2007.

Kipnis, Ken, Nancy M. P. King, and Robert M. Nelson. "An Open Letter to IRBs Considering Northfield Laboratories' PolyHeme Trial." *The American Journal of Bioethics* 6, no. 3 (May–June 2006): 1–4. Available online. URL: bioethics.net/journal/pdf/UAJB_A_166837.pdf. Accessed December 9, 2007.

Knowles, Francine. "News Socks Northfield Labs' Stock: Report Says Blood Substitute Resulted in Patient Deaths." *Chicago Sun-Times* (February 23, 2006).

Krugman, Saul. "The Willowbrook Hepatitis Studies Revisited: Ethical Aspects." *Review of Infectious Disease* 8 (January–February 1986): 157–162.

Lewin, Tamar. "U.S. Investigating Johns Hopkins Study of Lead Paint Hazard." *New York Times* (August 24, 2001). Available online. URL: query.nytimes.com/gst/fullpage.html?res=9405E1DF1331F937A1575BC0A9679C8B63. Accessed December 9, 2007.

Munson, Ronald, ed. "Research Ethics and Informed Consent." Chapter 1 in *Intervention and Reflection: Basic Issues in Medical Ethics.* 8th ed. Belmont, Calif.: Wadsworth/Cengage Learning, 2008.

Northfield Laboratories. "Northfield Laboratories Inc. Reports Fiscal 2007 Fourth Quarter and Year-End Financial Results." August 14, 2007. Available online. URL: www.fiercebiotech.com/node/8138/print. Accessed February 25, 2008.

Roig-Franzia, Manuel. "Probe Opens on Study Tied to Johns Hopkins." *Washington Post* (August 23, 2001). Available online. URL: www.washingtonpost.com/ac2/wp-dyn/A49123-2001Aug22?language=printer. Accessed December 9, 2007.

Rothman, David J., and Sheila M. Rothman. *The Willowbrook Wars.* New York: Harper and Row, 1984.

Steinbock, Bonnie, John D. Arras, and Alex John London, eds. "Experimentation on Human Subjects." Part 6 of *Ethical Issues in Modern Medicine: Contemporary Readings in Bioethics.* 7th ed. New York: McGraw-Hill, 2008.

Chapter 3: Research with Animals

Data on the use of animals in biomedical research comes largely from the U.S. Department of Agriculture and from the U.S. Humane Society's Web site. Information on Harry Harlow's psychological experiments on rhesus monkeys is drawn mainly from Deborah Blum's *The Monkey Wars* and *Love at Goon Park,* and from Peter Singer's *Animal Liberation.* For issues regarding xenotransplantation, the companion Web site for Frontline's *Organ Farm* is an important research tool. All sources are detailed below.

Blum, Deborah. *Love at Goon Park: Harry Harlow and the Science of Affection.* New York: Perseus Publishing, 2002.

———. *The Monkey Wars.* New York: Oxford University Press, 1994.

Cohen, Carl. "The Case for the Use of Animals in Biomedical Research." *New England Journal of Medicine* 315, no. 14 (October 2, 1986): 865–870.

Cook, Gareth. "OncoMouse Breeds Controversy: Cancer Researchers at Odds with DuPont over Fees for Patents." *San Francisco Chronicle* (June 3, 2002).

The Humane Society of the United States. "An Introduction to Primate Issues." Available online. URL: www.hsus.org/animals_in_research/general_information_on_animal_research/an_introduction_ to_primate_issues.html. Accessed January 2, 2008.

———. "Chimps Deserve Better Campaign Made Progress in 2007." January 19, 2008. Available online. URL: www.hsus.org/animals_in_research/chimps_deserve_better/chimps_deserve_better_1.html. Accessed March 1, 2008.

———. "Frequently Asked Questions about Chimpanzees in Research." Available online. URL: www.hsus.org/animals_in_research/chimps_deserve_better/chimpanzees_in_research_fact.html. Accessed January 2, 2008.

———. "The HSUS Praises Federal Agencies' Cooperation in Developing Non-Animal Toxicity Testing Methods." February 14, 2008. Available online. URL: www.hsus.org/press_and_publications/press_releases/the_hsus_praises_federal_cooperation_for_humane_testing_021408.html. Accessed March 1, 2008.

Nelson, Bryn. "Pig Livers to the Rescue?" *Newsday* (August 20, 2002). Available online. URL: sci.rutgers.edu/forum/archive/index.php/t-1115.html. Accessed January 2, 2008.

Pence, Gregory E. *Classic Cases in Medical Ethics: Accounts of Cases That Have Shaped Medical Ethics, with Philosophical, Legal, and Historical Backgrounds.* New York: McGraw-Hill, 2003.
Shorett, Peter. "Of Transgenic Mice and Men." *GeneWatch* 15, no. 5 (September 2002). Available online. URL: www.gene-watch.org/genewatch/articles/15-5mice.html. Accessed January 2, 2008.
Singer, Peter. *Animal Liberation.* 2d ed. New York: New York Review/Random House, 1990.
U.S. Department of Agriculture. "Animals Used in Research, 2006." Available online. URL: www.aphis.usda.gov/animal_welfare/downloads/awreports/awreport2006.pdf. Accessed February 25, 2008.
WGBH/Frontline, PBS. *Organ Farm.* Transcripts, interviews, and other materials available online. URL: www.pbs.org/wgbh/pages/frontline/shows/organfarm/. Accessed December 10, 2007.
World Intellectual Property Organization. "Bioethics and Patent Law: The Case of the OncoMouse." *WIPO Magazine* (May–June 2006). Available online. URL: www.wipo.int/wipo_magazine/en/2006/03/article_0006.html. Accessed January 2, 2008.

Chapter 4: When Life Ends

For the Nancy Cruzan story and the U.S. Supreme Court's landmark "right-to-die" ruling, the chapter "Euthanasia and Physician-Assisted Suicide" in Ronald Munson's *Intervention and Reflection: Basic Issues in Medical Ethics* is a key resource. Theresa Schiavo's story is drawn from a series of *New York Times* articles on her life and death, as well as published reflections of her guardian *ad litem*, Jay Wolfson. Historical background on defining death is provided largely by Ronald Munson's book *Raising the Dead*, by Steven Laureys's article, "Death, Unconsciousness and the Brain," and by the 1968 report of the Harvard Ad Hoc Committee. All sources are detailed below.

Ad Hoc Committee of the Harvard Medical School to Examine the Definition of Brain Death. "A Definition of Irreversible Coma." *Journal of the American Medical Association* 205, no. 6 (1968): 85–88.
Belluck, Pam. "Even as Doctors Say Enough, Families Fight to Prolong Life." *New York Times* (March 27, 2005). Available online. URL: www.nytimes.com/2005/03/27/national/27death.html. Accessed January 2, 2008.

Brody, Baruch A. "How Much of the Brain Must Be Dead?" In *Ethical Issues in Modern Medicine,* 6th ed. Edited by Bonnie Steinbock, John D. Arras, and Alex John London. New York: McGraw-Hill, 2003.

Carey, Benedict. "Inside the Injured Brain, Many Kinds of Awareness." *New York Times* (April 5, 2005). Available online. URL: www.nytimes.com/2005/04/05/health/05coma.html. Accessed January 2, 2008.

Carey, Benedict and John Schwartz. "Schiavo's Condition Holds Little Chance of Recovery." *New York Times* (March 26, 2005). Available online. URL: www.nytimes.com/2005/03/26/health/26brain.html. Accessed January 10, 2008.

Chiong, Winston. "Brain Death without Definitions." *Hastings Center Report* 35, no. 6 (November–December 2005): 20–30.

Fagerlin, Angela, and Carl E. Schneider. "Enough: The Failure of the Living Will." *Hastings Center Report* 34, no. 2 (March–April 2004): 30–42.

Goodnough, Abby. "Schiavo Dies, Ending Bitter Case Over Feeding Tube." *New York Times* (April 1, 2005). Available online. URL: www.nytimes.com/2005/04/01/national/01schiavo.html. Accessed January 2, 2008.

———. "Schiavo Autopsy Says Brain, Withered, Was Untreatable." *The New York Times* (June 16, 2005). Available online. URL: www.nytimes.com/2005/06/16/national/16schiavo.html. Accessed January 2, 2008.

Goodnough, Abby and William Yardley. "Federal Judge Condemns Intervention in Schiavo Case." *New York Times* (March 31, 2005). Available online. URL: www.nytimes.com/2005/03/31/national/31schiavo.html. Accessed January 2, 2008.

Hershenov, David. "The Problematic Role of 'Irreversibility' in the Definition of Death." *Bioethics* 17, no. 1 (February 2003): 89–100.

Laureys, Steven. "Death, Unconsciousness and the Brain." *Nature Reviews Neuroscience* 6, no. 11 (November 2005): 899–909. Available online. URL: www.coma.ulg.ac.be/papers/vs/death_unconsciousness_NatureRevNeurosci05.pdf. Accessed January 10, 2008.

Lock, Margaret. *Twice Dead: Organ Transplants and the Reinvention of Death.* Berkeley, Calif.: University of California Press, 2001.

McKinley, Jesse. "Surgeon Accused of Speeding a Death to Get Organs." *New York Times* (February 27, 2008). Available online.

URL: www.nytimes.com/2008/02/27/us/27transplant.html?_r=1&oref=slogin. Accessed February 29, 2008.

Munson, Ronald, ed. "Euthanasia and Physician-Assisted Suicide." Chapter 11 in *Intervention and Reflection: Basic Issues in Medical Ethics*. 8th ed. Belmont, Calif.: Wadsworth/Cengage Learning, 2008.

Munson, Ronald. *Raising the Dead: Organ Transplants, Ethics, and Society*. New York: Oxford University Press, 2001.

"National Briefing/South/Florida: Poll on Schiavo Case." *New York Times* (April 14, 2005). Available online. URL: query.nytimes.com/gst/fullpage.html?res=9A02E6DE103EF937A25757C0A9639C8B63. Accessed January 2, 2008.

Payne, Kirk, Robert M. Taylor, Carol Stocking, and Greg A. Sachs. "Physicians' Attitudes about the Care of Patients in the Persistent Vegetative State: A National Survey." *Annals of Internal Medicine* 125, no. 2 (July 1996): 104–110. Available online. URL: www.annals.org/cgi/content/full/125/2/104. Accessed January 10, 2008.

Schwartz, John. "Neither 'Starvation' nor the Suffering It Connotes Applies to Schiavo, Doctors Say." *New York Times* (March 25, 2005). Available online. URL: www.nytimes.com/2005/03/25/national/25starve.html. Accessed January 2, 2008.

Schwartz, John and James Estrin. "Many Still Seek One Final Say on Ending Life." *New York Times* (June 17, 2005). Available online. URL: www.nytimes.com/2005/06/17/health/17will.html. Accessed January 2, 2008.

Selzer, Richard. *Raising the Dead: A Doctor's Encounter with His Own Mortality*. East Lansing, Mich.: Michigan State University Press, 2001.

Siminoff, Laura A., Christopher Burant, and Stuart J. Youngner. "Death and Organ Procurement: Public Beliefs and Attitudes." *Kennedy Institute of Ethics Journal* 14, no. 3 (September 2004). 217–234.

Stolberg, Sheryl Gay. "Schiavo's Case May Reshape American Law." *New York Times* (April 1, 2005). Available online. URL: www.nytimes.com/2005/04/01/politics/01legacy.html?adxnnl=1&adxnnlx=1199300724-HS4FrhbJHXRzCCpIQBgXhQ. Accessed January 2, 2008.

Wolfson, Jay. "Erring on the Side of Theresa Schiavo: Reflections of the Special Guardian ad Litem." *Hastings Center Report* 35, no. 3 (May–June 2005): 16–19.

Chapter 5: Life-Extending Technology

Information on the Tirhas Habtegiris story comes from local (Dallas, Texas) news reports, as well as from media materials from the Baylor Health Care System; the English car-accident victim's story from investigators' reports (and responses to them) in *Science* magazine and from secondary media sources; and the Ayalas' story from articles in the *New York Times* and *Time* magazine. All sources are detailed below.

"Baby Is Conceived to Save Daughter." Associated Press (February 17, 1990). Available online. URL: query.nytimes.com/gst/fullpage.html?res=9C0CE6D71630F934A25751C0A966958260. Accessed January 2, 2008.

Carey, Benedict. "Mental Activity Seen in a Brain Gravely Injured." *New York Times* (September 8, 2006). Available online. URL: www.nytimes.com/2006/09/08/science/08brain.html?_r=1&oref=slogin. Accessed January 11, 2008.

"Conceived to Save Her Sister, a Child Is Born." Associated Press (April 7, 1990). Available online. URL: query.nytimes.com/gst/fullpage.html?res=9C0CE3DB1E31F934A35757C0A966958260. Accessed January 2, 2008.

Fink, Jack. "Family Debates Hospital Action in Woman's Death." CBS 11 (Dallas/Fort Worth) (December 14, 2005).

Fritz, Mark. "How Feeding Tube Figures into End-of-Life Debate." *Wall Street Journal* (December 8, 2005). Available online. URL: www.post-gazette.com/pg/05342/619286.stm. Accessed January 2, 2008.

Jordan, V. Craig. "Is Tamoxifen the Rosetta Stone for Breast Cancer?" *Journal of the National Cancer Institute* 95, no. 5 (March 5, 2003): 338–340. Available online. URL: jnci.oxfordjournals.org/cgi/content/full/95/5/338. Accessed January 14, 2008.

———. "Tamoxifen (ICI46,474) as a Targeted Therapy to Treat and Prevent Breast Cancer." *British Journal of Pharmacology* 147 (2006): S269–S276. Available online. URL: www.nature.com/bjp/journal/v147/n1s/full/0706399a.html. Accessed January 14, 2008.

Jordan, V. Craig and Monica Morrow. "Chemoprevention of Breast Cancer: A Model for Change." *Journal of Clinical Oncology* 20, no. 1 (January 1, 2002): 1–3. Available online. URL: jco.ascopubs.org/cgi/content/full/20/1/1. Accessed January 14, 2008.

Kolata, Gina. "Hormones and Cancer: Assessing the Risks." *New York Times* (December 26, 2006). Available online. URL: www.

nytimes.com/2006/12/26/health/26horm.html. Accessed January 14, 2008.

"A Life-Saving Sibling Helps Out Once More." *New York Times* (June 7, 1992). Available online. URL: query.nytimes.com/gst/fullpage.html?res=9E0CEFDA133DF934A35755C0A964958260. Accessed January 2, 2008.

Mayo Foundation for Medical Education and Research. "Organ Donation: Don't Let These 10 Myths Confuse You." April 4, 2008. Available online. URL: www.mayoclinic.com/health/organ-donation/FL00077. Accessed September 18, 2008.

Morrow, Lance et al. "When One Body Can Save Another." *Time* (June 17, 1991). Available online. URL: www.time.com/time/magazine/article/0,9171,973182,00.html. Accessed January 2, 2007.

Naccache, Lionel. "Is She Conscious?" *Science* 313 (September 8, 2006): 1,395–1,396. Available online. URL: www.mrc-cbu.cam.ac.uk/~adrian/Site/Newslist_files/Naccache_Science_Brevia_Comment.pdf. Accessed January 11, 2008.

Owen, Adrian M., Martin R. Coleman, Melanie Boly, Matthew H. Davis, Steven Laureys, and John D. Pickard. "Detecting Awareness in the Vegetative State." *Science* 313 (September 8, 2006): 1,402. Available online. URL: www.sciencemag.org/cgi/content/full/sci;313/5792/1402. Accessed January 11, 2008.

Owen, Adrian M., Martin R. Coleman, Melanie Boly, Matthew H. Davis, Steven Laureys, Dietsje Jolles, and John D. Pickard. "Response to Comments on 'Detecting Awareness in the Vegetative State.'" *Science* 315 (March 2, 2007): 1,221c. Available online. URL: www.sciencemag.org/cgi/content/full/315/5816/1221c. Accessed January 11, 2008.

Roche, Vivyenne. "Percutaneous Endoscopic Gastrostomy: Clinical Care of PEG Tubes in Older Adults." *Geriatrics* 58, no. 11 (November 2003): 22–26, 28–29.

Saletan, William. "The Unspeakable: Buried Alive in Your Own Skull." *Slate* (September 12, 2006). Available online. URL: www.slate.com/id/2149182/. Accessed January 11, 2008.

Stein, Rob. "'Vegetative' Woman's Brain Shows Surprising Activity." *Washington Post* (September 8, 2006). Available online. URL: www.washingtonpost.com/wp-dyn/content/article/2006/09/07/AR2006090700978.html. Accessed January 11, 2008.

St. James, Janet. "Woman's Death Highlights Health Insurance Crisis." WFAA-TV, Dallas Fort Worth (December 14, 2005). Available online.

URL: www.wfaa.com/sharedcontent/dws/wfaa/latestnews/stories/wfaa051214_lj_african.bb0e76d.html. Accessed January 2, 2008.

United Network for Organ Sharing. "Fact Sheets." Available online. URL: www.unos.org/inTheNews/factSheets.asp. Accessed January 2, 2008.

Chapter 6: Life Extension, Aging, and Palliative Care

Data on life expectancy is drawn primarily from the World Health Organization's *World Health Report* (2006 and 2007), as well as from Christopher J. L. Murray et al.'s study, "Eight Americas." AARP is a key resource for information on aging in America, and Stephen G. Post and Robert H. Binstock's *The Fountain of Youth: Cultural, Scientific and Ethical Perspectives on a Biomedical Goal* provides a variety of viewpoints on extending the human life span. All sources are detailed below.

AARP. "Long-Term Care: Public Perceptions Versus Reality in 2006." December 2006. Available online. URL: www.aarp.org/research/longtermcare/costs/ltc_costs_fs_2006.html. Accessed January 3, 2008.

———. "In Brief: A Growing Crisis in Health and Long-Term Services and Supports for Older Persons with Disabilities: Changes from 2002–2005." December 2006. Available online. URL: www.aarp.org/research/housing-mobility/homecare/inb133_il.html. Accessed January 3, 2008.

Bostrom, Nick. "Recent Developments in the Ethics, Science, and Politics of Life Extension." *Ageing Horizons* no. 3 (Autumn–Winter 2005): 28–33. Available online. URL: www.ageing.ox.ac.uk/ageinghorizons/thematic%20issues/biodemography/papers%20biodemography/pdf/4bostromah3.pdf. Accessed January 3, 2008.

Brown, David. "Wide Gaps Found In Mortality Rates Among U.S. Groups." *Washington Post* (September 12, 2006). Available online. URL: www.washingtonpost.com/wp-dyn/content/article/2006/09/11/AR2006091101297.html. Accessed January 3, 2008.

Fleck, Carol. "Double Bind: As Boomers Juggle Work and Caring for Their Aging Parents, Business Also Pays a Price." *AARP Bulletin* (May 2006). Available online. URL: www.aarp.org/family/caregiving/articles/cost_elder_care.html. Accessed January 3, 2008.

Genworth Financial. "2008 Cost of Care Survey." April 2008. Available online. URL: www.genworth.com/content/etc/medialib/genworth/us/en/Long_Term_Care.Par.14291.file/dat/37522%20CoC%20Brochure.pdf. Accessed September 18, 2008.

Green, Jeff. "U.S. Has Second Worst Newborn Death Rate in Modern World, Report Says." CNN.com (May 10, 2006). Available online. URL: www.cnn.com/2006/HEALTH/parenting/05/08/mothers.index/index.html. Accessed January 3, 2008.

Gross, Jane. "New Options (and Risks) in Home Care for Elderly" (March 1, 2007) Available online. URL: www.nytimes.com/2007/03/01/us/01aides.html?_r=1&adxnnl=1&oref=slogin&adxnnlx=1199386919-MAWmjl3Ff4X867jI8 V3/VA. Accessed January 3, 2008.

Hospice Foundation of America. "What Is Hospice?" Available online. URL: www.hospicefoundation.org/hospiceInfo. Accessed January 3, 2008.

Kass, Leon R. *Toward a More Natural Science: Biology and Human Affairs*. New York: Free Press, 1988.

Kolata, Gina. "Pushing Limits of the Human Life Span." *New York Times* (March 9, 1999). Available online. URL: query.nytimes.com/gst/fullpage.html?res=9D03E2DB103FF93AA35750C0A96F958260. Accessed January 3, 2008.

Murray, Christopher J. L., Sandeep C. Kulkarni, Catherine Michaud, Niels Tomijima, Maria T. Bulzacchelli, Terrell J. Landiorio, Majid Ezzati. "Eight Americas: Investigating Mortality Disparities across Races, Counties, and Race-Counties in the United States." *PLoS Medicine* 3, no. 9 (September 2006): 1,513–1,524. Available online. URL: medicine.plosjournals.org/perlserv/?request=get-document&doi=10.1371/journal.pmed.0030260. Accessed January 3, 2008.

National Institutes of Health. "NIH Profile: Cynthia Kenyon, PhD." April 26, 2005. Available online. URL: www.nih.gov/about/people/kenyon.htm. Accessed January 3, 2008.

Novotny, Thomas E. "Why We Need to Rethink the Diseases of Affluence." *PLoS Medicine* 2, no. 5 (May 2005): e104. Available online. URL: medicine.plosjournals.org/perlserv?request=get-document&doi=10.1371/journal.pmed.0020104. Accessed January 12, 2008.

Post, Stephen G., and Robert H. Binstock, eds. *The Fountain of Youth: Cultural, Scientific and Ethical Perspectives on a Biomedical Goal.* New York: Oxford University Press, 2004.

The President's Council on Bioethics. *Beyond Therapy: Biotechnology and the Pursuit of Happiness.* New York: Harper Perennial, 2003. Available online. URL: www.bioethics.gov/reports/beyondtherapy. Accessed January 3, 2008.

"Suing to Get Out in the World: San Franciscans Break Out of Nursing Homes to Live on Their Own." *AARP Bulletin* (June 2004). Available online. URL: www.aarp.org/family/caregiving/article/suing.html. Accessed January 3, 2008.

Swafford, Stephanie. "Diseases of Affluence Hit Developing Nations." *BMJ* 314 (May 10, 1997): 1,365. Available online. URL: www.bmj.com/cgi/content/full/314/7091/1365/d. Accessed January 12, 2008.

"Will You Still Feed Me?" *New York Times* (April 28, 2002). Available online. URL: query.nytimes.com/gst/fullpage.html?res=9B03E2DB1E3EF93BA15757C0A9649C8B63. Accessed January 3, 2008.

World Health Organization. "The World Health Report 1997: Conquering Suffering, Enriching Humanity." Available online. URL: www.who.int/whr/previous/en. Accessed January 12, 2008.

———. "Rethinking 'Diseases of Affluence': The Economic Impact of Chronic Diseases." Facing the Facts #4, 2005. Available online. URL: www.who.int/chp/chronic_disease_report/media/Factsheet4.pdf. Accessed January 12, 2008.

———. "The World Health Report 2006: Working Together for Health." Available online. URL: www.who.int/whr/2006/en/index.html. Accessed January 12, 2008.

———. "World Health Statistics 2006." Available online. URL: www.who.int/whosis/whostat2006/en. Accessed January 12, 2008.

Chapter 7: New Technology and the Cost of Treatment

Information on rising health care costs is drawn from a wide variety of sources, including national and international government documents, recent press reports, and surveys/studies conducted by nonprofit organizations and private foundations concerned with the equitable distribution of quality health care. For information on the HIV virus and its prevention and treatment, the companion Web site for Frontline's *The Age of AIDS* is a valuable research tool. All sources are detailed below.

Abramson, John. *Overdo$ed America: The Broken Promise of American Medicine.* New York: Harper Perennial, 2005.

Angell, Marcia. *The Truth About the Drug Companies: How They Deceive Us and What to Do About It*. New York: Random House, 2004.
Berenson, Alex. "Pinning Down the Money Value of a Person's Life." *New York Times* (June 11, 2007). Available online. URL: www.nytimes.com/2007/06/11/business/businessspecial3/11life.html?_r=1&o ref=slogin. Accessed February 4, 2008.
Callahan, Daniel. *False Hopes: Why America's Quest for Perfect Health is a Recipe for Failure*. New York: Simon & Schuster, 1998.
Commonwealth Fund. "Mirror, Mirror on the Wall: An International Update on the Comparative Performance of American Health Care," May 15, 2007. Available online. URL: www.commonwealthfund.org/publications/publications_show.htm?doc_id=482678. Accessed February 6, 2008.
Gladwell, Malcolm. "High Prices: How to Think About Prescription Drugs." *New Yorker* (October 25, 2004). Available online. URL: www.newyorker.com/archive/2004/10/25/041025crat_atlarge. Accessed February 4, 2008.
Harris, Gardiner. "Prilosec's Maker Switches Users To Nexium, Thwarting Generics." *Wall Street Journal* (June 6, 2002).
Institute for OneWorld Health. "The Global Burden of Infectious Disease." Available online. URL: www.oneworldhealth.org/global/global_burden.php. Accessed February 6, 2008.
Kaiser Family Foundation. *Employer Health Benefits 2005 Annual Survey*. Available online. URL: www.kff.org/insurance/7315.cfm. Accessed February 6, 2007.
McNeil, Donald G., Jr. "U.S. Urges HIV Tests for Adults and Teenagers." *New York Times* (September 22, 2006). Available online. URL: www.nytimes.com/2006/09/22/health/22hiv.html?adxnnl=1&adxnnlx=1202140850 jLjf6VJXQWlr2ye9ymFLWQ. Accessed February 4, 2008.
National Coalition on Health Care. "Facts on the Cost of Health Care." Available online. URL: www.nchc.org/facts/cost.shtml. Accessed February 4, 2008.
Organisation for Economic Co-operation and Development (OECD). "Press Release: OECD Health Data 2006," June 26, 2006. Available online. www.bfs.admin.ch/bfs/portal/en/index/news/medienmitteilungen.Document.78573.pdf. Accessed February 4, 2008.
Pear, Robert. "Child Health Care Splits White House and States." *New York Times* (February 27, 2007). Available online. URL: www.

nytimes.com/2007/02/27/washington/27govs.html. Accessed February 4, 2008.

State Health Access Data Assistance Center (SHADAC) and the Urban Institute, prepared for the Robert Wood Johnson Foundation. "Going Without: America's Uninsured Children," August 2005. Available online. URL: www.rwjf.org/files/newsroom/ckfresearchreportfinal.pdf. Accessed February 4, 2008.

Stolberg, Sheryl Gay. "President Vetoes Second Measure to Expand Children's Health Program." *New York Times* (December 13, 2007). Available online. URL: www.nytimes.com/2007/12/13/us/13bush.html?_r=1&oref=slogin. Accessed November 4, 2008.

Toner, Robin and Janet Elder. "Poll Shows Majority Back Health Care for All." *New York Times* (March 1, 2007). Available online. URL: www.nytimes.com/2007/03/01/washington/01cnd-poll.html?ex=1330405200&en=45c0a4cf48ed21a1&ei=5088&partner=rssnyt&emc=rss. Accessed February 4, 2008.

Tong, Rosemary. *New Perspectives in Healthcare Ethics: An Interdisciplinary and Crosscultural Approach.* Upper Saddle River, N.J.: Prentice Hall, 2006.

UNAID. "Sub-Saharan Africa: AIDS Epidemic Update Regional Summary." March 2008. Available online. URL: data.unaids.org/pub/Report/2008/jc1526_epibriefs_ssafrica_en.pdf. Accessed October 22, 2008.

U.S. Department of Health and Human Services. Office of the Assistant Secretary for Planning and Evaluation. "Overview of the Uninsured in the United States: An Analysis of the 2007 Current Population Survey," September 2007. Available online. URL: aspe.hhs.gov/health/reports/07/uninsured/index.htm. Accessed February 4, 2008.

WGBH/Frontline, PBS. *The Age of AIDS.* Transcripts, interviews, and other materials available online. URL: www.pbs.org/wgbh/pages/frontline/aids/. Accessed April 30, 2008.

Chapter 8: Health, Disease, and Wellness

Information on liver transplants for alcoholics and on mental health parity comes largely from materials by the National Institute on Alcohol Abuse and Alcoholism and by the National Institute of Mental Health, and from recent press reports. Background on the human enhancement debate is drawn from Erik Parens's article, "Authenticity and Ambivalence," as well as from recent press reports and the World Transhumanist Association Web site. All sources are detailed below.

Alexander, Brian. "Is There a Right to be Superhuman?" MSNBC. com. Available online. URL: www.msnbc.msn.com/id/13054181. Accessed January 15, 2008.

American Psychological Association (APA). "APA Poll: Most Americans Have Sought Mental Health Treatment but Cost, Insurance Still Barriers," May 13, 2004. Available online. URL: www.apa.org/releases/practicepoll_04.html. Accessed January 15, 2008.

———. "Answers to Your Questions for a Better Understanding of Sexual Orientation and Homosexuality." Available online. URL: www.apa.org/topics/sorientation.html. Accessed September 19, 2008.

APA Online. "Landmark Victory: Mental Health Parity Is Now Law." October 3, 2008. Available online. URL: www.apapractice.org/apo/in_the_news/parity_passes_press.html#. Accessed November 4, 2008.

Byers, J. Wellington. "Diseases of the Southern Negro." *Medical and Surgical Reporter* 43 (1888): 735.

International Labour Organization. "Mental Health in the Workplace." Available online. URL: www.ilo.org/public/english/employment/skills/disability/papers/execsumcontents.htm. Accessed January 15, 2008.

Longman, Jeré. "An Amputee Sprinter: Is He Disabled or Too-Abled?" *New York Times* (May 15, 2007). Available online. URL: www.nytimes.com/2007/05/15/sports/othersports/15runner.html?_r=1&oref=slogin. Accessed January 15, 2008.

Parens, Erik. "Authenticity and Ambivalence: Toward Understanding the Enhancement Debate." *Hastings Center Report* 35, no. 3 (May–June 2005): 34–41.

Pear, Robert. "Study Backs Equal Coverage for Mental Illness." *New York Times* (March 30, 2006). Available online. URL: www.nytimes.com/2006/03/30/health/30mental.html?pagewanted=print. Accessed January 15, 2007.

Rovner, Julie. "Mental Health Parity Approved with Bailout Bill." National Public Radio. October 6, 2008. Available online. URL: www.npr.org/templates/story/story.php?storyId=95435676 Accessed November 4, 2008.

Saletan, William. "Among the Transhumanists: Cyborgs, Self-Mutilators, and the Future of Our Race." *Slate* (June 4, 2006). Available online. URL: www.slate.com/id/2142987/. Accessed January 15, 2008.

U.S. Department of Health and Human Services (HHS), National Institutes of Health (NIH). National Institute on Alcohol Abuse and Alcoholism (NIAAA). "Alcoholic Liver Disease." *Alcohol Alert* 64 (January 2005). Available online. URL: pubs.niaaa.nih.gov/publications/aa64/aa64.htm. Accessed January 15, 2008.

———. National Institute of Mental Health (NIMH). "The Numbers Count: Mental Disorders in America." Available online. URL: www.nimh.nih.gov/health/publications/the-numbers-count-mental-disorders-in-america.shtml. Accessed January 15, 2008.

U.S. Department of Labor. Bureau of Labor Statistics. "New Law Moves Insurance Plans Closer to Total Mental Health Parity," September 22, 2003. Available online. URL: www.bls.gov/opub/cwc/cm20030909ar01p1.htm. Accessed January 15, 2008.

World Transhumanist Association. "The Transhumanist Declaration," March 4, 2002. Available online. URL: transhumanism.org/index.php/WTA/declaration. Accessed January 15, 2008.

Web Sites

For the latest information on the medical research and treatment topics considered in this volume, readers can consult the Web sites of the following government agencies, associations, nonprofit organizations, and professional journals. All provide searchable text online and may also provide interactive features, video clips, podcasts, and links to other relevant sources.

AARP (formerly the American Association of Retired Persons). URL: www.aarp.org. Accessed April 27, 2008. Information on a range of topics related to aging.

American Association for the Advancement of Science (AAAS). URL: www.aaas.org. Accessed April 27, 2008. News related to a broad range of scientific topics and careers.

American Journal of Bioethics. URL: www.bioethics.net. Accessed April 27, 2008. Free access to abstracts and some full text articles, online discussions, and links to relevant news articles.

American Psychological Association (APA). URL: www.apa.org. Accessed April 27, 2008. Information on issues and careers in psychology.

Centers for Disease Control and Prevention (CDC). URL: www.cdc.gov. Accessed April 27, 2008. Information on a wide range of health and safety topics.

Further Resources

Hospice Foundation of America (HFA). URL: www.hospicefoundation.org. Accessed April 27, 2008. Information and support resources for people living with terminal diseases and their families and friends.

Humane Society of the United States (HSUS). URL: www.hsus.org. Accessed April 27, 2008. Information about the treatment and protection of animals in the United States, including animals used in medical research.

Institute for OneWorld Health (iOWH). URL: www.oneworldhealth.org/about/index.php. Accessed April 27, 2008. Information on global diseases and the search for cures, from the first nonprofit pharmaceutical company in the United States.

Mayo Foundation for Medical Education and Research. URL: www.mayoclinic.com. Accessed April 27, 2008. Medical information and online resources from the Mayo Clinic.

National Cancer Institute (NCI). URL: www.cancer.gov. Accessed April 27, 2008. Cancer-related information of all kinds, including descriptions of NCI research programs and clinical trials.

National Institute of Allergy and Infectious Diseases (NIAID). URL: www3.niaid.nih.gov. Accessed April 27, 2008. Information on a vast array of infectious diseases and their prevention and treatment.

National Institute of Mental Health (NIMH). URL: www.nimh.nih.gov. Accessed April 27, 2008. News and information on a range of issues related to mental health.

National Institutes of Health (NIH). URL: www.nih.gov. Accessed April 27, 2008. News and information on a wide range of medical research, funding, and career topics.

Public Library of Science (PLoS). URL: www.plos.org. Accessed April 27, 2008. Open access to articles on a wide range of topics in the medical and biological sciences.

State Children's Health Insurance Program (SCHIP). URL: www.cms.hhs.gov/home/schip.asp. Accessed April 27, 2008. Information on low-cost medical insurance for families and children.

United Network for Organ Sharing (UNOS). URL: www.unos.org. Accessed April 27, 2008. Up-to-date data on organ transplants and available donors.

U.S. Department of Health and Human Services (HHS). URL: www.hhs.gov. Accessed April 27, 2008. Links to information on a variety of health and medical ethics issues.

World Health Organization (WHO). URL: www.who.int/en. Accessed April 27, 2008. Information on a wide range of global health topics from the coordinating authority for health within the United Nations system.

World Medical Association (WMA). URL: www.wma.net. Accessed April 27, 2008. Information from the global representative body for physicians.

Free Online Print and Radio Media

Several of the print and radio news sources referenced in this volume are available free to online users. The following Web sites contain bonus audiovisual content, such as video clips, slide shows, interactive graphics, and podcasts.

National Public Radio (NPR). www.npr.org. Accessed April 27, 2008.
Newsday. URL: www.newsday.com. Accessed April 27, 2008.
Newsweek. URL: www.newsweek.com. Accessed April 27, 2008.
New York Times. URL: www.nytimes.com. Accessed April 27, 2008.
San Francisco Chronicle. URL: www.sfgate.com. Accessed April 27, 2008.
Slate. URL: www.slate.com. Accessed April 27, 2008.
Time. URL: www.time.com. Accessed April 27, 2008.
Washington Post. URL: www.washingtonpost.com. Accessed April 27, 2008.

INDEX

Note: *Italic* page numbers indicate illustrations. Page numbers followed by *t* denote tables, charts, or graphs; page numbers followed by *m* denote maps.

A

abatement 33
ABC (Abstinence, Be faithful, and use Condoms) 154, 156
Abramson, John 142, 144–145
abstinence-only programs 148, 149, 156
active control trial 20
active infection 68
acute rejection 67
Advance Directives Act (Texas Futile Care Law) 101, 104, 179*c*
Africa 15–20, 126, 147, 148
African Americans
    19th-century medical attitudes towards 159–160
    life expectancy 126–127
    Tuskegee syphilis study 1–8, *4*, *7*
aging 122–135
Aging with Dignity 91
Agriculture, U.S. Department of 47, 61
AIDS. *See* HIV/AIDS pandemic
alcoholic liver disease (ALD) 161–162
alcoholics, liver transplants for 161–163
alcohol-related end-stage liver disease (ARESLD) 162, 163
Allan, Jonathan 69
American Journal of Bioethics 42–43

Andersen, Hans Christian 76
Anderson, Walter Truett 172
Angell, Marcia 20–21, 145
Animal Liberation (Singer) 51–52, 59–60, 175*c*
animal research 46–73, 47*t*–49*t*
    ethical positions on 50–54
    Harry Harlow and attachment theory 54, *58*, 58–61
    protections for animal subjects 61–63
    and xenotransplantation 63–72, *65*, *71*
animal rights 52–53
animal suffering 50–51, 70, *71*, 72–73
Animal Welfare Act 61, 175*c*, 176*c*
anxiety disorders 164–165
ARESLD (alcohol-related end-stage liver disease) 162, 163
artificial respirator. *See* mechanical ventilator
assent 28
AstraZeneca 102, 142
attachment theory 54, *58*, 58–61
Auchincloss, Hugh, Jr. 68–69
Auerbach, Judith 151
Austin, Amanda L. 167
Avner, Dennis 169
Ayala, Abe 119
Ayala, Anissa *118*, 118–120, 178*c*
Ayala, Marissa Eve *118*, 120, 178*c*
Ayala, Mary 119
AZT (ZDV) 16–17, 20, 152

B

Bach, Fritz 69, 70
Barnard, Christiaan 177*c*

213

Beecher, Henry K. 82
  ethics of organ donation 82–83
  on irreversible coma 101
  on Willowbrook study 26, 174c–175c
Belmont Report
  ethical principles outlined in 11, 14, 45
  HIV vertical transmission trials 21, 22
  release of 11, 175c
  Tuskegee syphilis study 6
beneficence 12, 14, 19–21
Bentham, Jeremy 52
Berger, Alan 69
Bernat, James L. 79
best interests standard 93–94
Beyrer, Chris 149
Biederman, Joseph 171
biogerontology 122–135
Biopure 44
bipolar disorder 164
black fever (visceral leishmaniasis) 145
Blackwell, Elizabeth 159
Blum, Deborah 60
bone marrow transplantation 118–120, 178c
Bostrom, Nick 126
brain damage 108
brain death
  as clinical criterion for death 79
  and heart-lung death 84
  and organ donation 117
  PVS v. 83
  Karen Quinlan case 75
  Theresa Schiavo case 88–89
brain stem 89, 176c
Brandt, Karl 8
Brazil 156
breast cancer 102–104, *103*
Brody, Baruch A. 80
Brown, William J. 2
Brutoco, Rudolf 120
bulimia 88
Burch, Rex 62
Burger, Warren 55

Bush, George W.
  Advance Directives Act 104
  AIDS policies 148, 156
  children's health insurance bill vetoes 140
  human enhancement policies 171
  Theresa Schiavo case 95, 96
Bush, Jeb 95, 96
Buxton, Peter 2
Byers, J. Wellington 160

C

Cade, Ebb 36–37, 174c
Caenorhabditis elegans 124
Callahan, Daniel 141–142
Campisi, Judith 124
Canada 57
cancer
  Tirhas Habtegiris case 101–106
  OncoMouse 55–57, *56*
  plutonium injection experiments 35
  tamoxifen and 102–104, *103*
Capron, Alexander Morgan 119
Case for Animal Rights, The (Regan) 52–53
Cathell, Dale R. 31–32
Catholic Church. *See* Roman Catholicism
cats 49–50
CAT (computerized axial tomography) scan 109–110, 177c
CD4 cell 151, 152
centenarians 128
Chapman, Audrey 125
Charo, R. Alta 96
children
  conception for organ transplants *118*, 118–120
  drugs to treat behavior 170–171
  health insurance for 139–140
  KKI lead-based paint study 29–32, *30*, 175c, 176c
  protections for 27–29
  Willowbrook hepatitis study 25–27

Index

Childs, Nancy 89
chimeras 66
chimpanzees 48, 67, 174c
chronic disease 129–131, 130t, 131t
cirrhosis 161, 162
clinical depression (major depressive disorder) 59
clinical trial 104
Clinton, Bill 7, 8, 35, 167, 175c
coercion 19
Cohen, Carl 53
cold war 34
coma
   and brain death 82–83
   Cruzan v. Missouri 108
   defined 89
   and mechanical ventilator 100
   Karen Ann Quinlan case 75, 177c
computerized axial tomography (CT/CAT) scan 109–110, 177c
condoms 154, 156
Congress, U.S. 95, 140, 149, 163–164
consciousness 83
consent 35, 36, 174c
control group 6, 17
Cormack, Allan McLeod 109, 177c
Corzine, Jon 140
costs of health care
   cost-effectiveness determination 141t
   drug expenditure per capita, OECD countries, 2004 143t
   influence of pharmaceutical gifts on prescribing practices 144t
   national expenditures 136–138, 137t, 138t
   technology and 140–146
Cranford, Ronald 90
critical function 79
Cruzan, Nancy 108, 177c, 178c
Cruzan v. Missouri 78, 108–109, 178c
CT (computerized axial tomography) scan 109–110, 177c

cultural assumptions 160, 161
cyclosporine 81, 115, 177c

D

DAF2 (decay accelerating factor) gene 124
Daniels, Norman 69, 70
death
   clinical criteria 79–81, 80t
   defining 76–85, 77, 78, 80t
   ethical issues. *See* end-of-life issues
   hastening of v. extending life 85–87
   heart-lung criteria 84–85
Declaration of Helsinki 10, 20, 174c
deoxyribonucleic acid (DNA) 151
depressive disorders 164
direct benefit 38
disease 158–160
DNA (deoxyribonucleic acid) 151
Doctor's Trial 8–10, 9
dogs 49
Domenici, Pete 167
Douglas, Jim 140
Draize, John H. 50
Draize test 50
Dresser, Rebecca 92–94
drug resistance 104
drugs. *See also specific drugs or classes of drugs, e.g.*: AZT, immunosuppressive drugs
   costs of research, development, and delivery 140–146
   to treat childhood behavior 170–171
Dubos, René 168
DuPont (chemical company) 57

E

eating disorders 88
Ehrlich, Paul 102
"Eight Americas" 127
electrocardiogram (EKG/ECG) 81
electroencephalogram 90
embryonic stem cells 117

emergency room patients 34–45, 40, 175c
encapsulation 33
endemic disease 25
end-of-life issues 74–97
   Cruzan v. Missouri 108–109
   defining death 76–85, 77, 78, 80t
   living wills and medical proxies 91–94
   Karen Quinlan case 74–76
   Theresa Schiavo case 87–90, 94–96
endogenous retroviruses 68
endothelial cells 64, 66
endotracheal tube 99
enhancement, health v. 168–172
"Enough: The Failure of the Living Will" (Schneider and Fagerlin) 93
enteral feeding. *See* feeding tube
entry inhibitors 152
Environmental Protection Agency, U.S. (EPA) 29, 34
EPO (European Patent Office) 55–56
ER (estrogen receptor)-positive tumor 104
ethical principles in medical research 1–22
   animal research 50–54
   Belmont Report 11, 14
   and cultural assumptions 161
   Declaration of Helsinki 10
   defining health and disease 158–160
   end-of-life issues 74–97
   equal treatment for mental illness 163–168, 165t, 167
   HIV vertical transmission trials 15–22, 16
   human life extension 125–126
   Institutional Review Boards (IRBs) 11–13
   liver transplants for alcoholics 161–163
   Nuremberg Code 8–10, 9
   Tuskegee syphilis study 1–8, 4, 7
   U.S. regulations on protection of human subjects 10–11
   vulnerable populations 23–45
"Ethics and Clinical Research" (Beecher) 26
European Patent Office (EPO) 55–56
extraordinary care 87, 176c
extubation 106

F

Fagerlin, Angela 93
Fairman, Ronald M. 41
Fauci, Anthony 151–152
FDA. *See* Food and Drug Administration, U.S.
feeding tube
   ethical issues 106–109
   ethical problems 77
   invention of 177c
   Theresa Schiavo case 88, 95, 178c
Finucane, Thomas 107
fMRI (functional MRI) 110–112, 114, 179c
Food and Drug Administration, U.S. (FDA)
   PolyHeme tests 39, 41–45
   tamoxifen approval 104
   and xenotransplantation 67, 68
Forrow, Lachlan 96
Fost, Norman 86–87
Frank, Richard G. 167
functional MRI. *See* fMRI
fusion inhibitors 153
futile treatment 96

G

GAL (galactose-alpha,1,3-galactose) 64, 66
gamma globulin 25
Gan, T. J. 41
Gauderer, Michael 107
Gayle, Helene 21
GDP (gross domestic product) 136, 138

Index

genetic engineering 55–57, *56*, 63, 123–124
Giles, Joan P. 25
Goldby, Stephen 26–27
Goldkind, Sara 44–45
Goldstein, Gary 32
Gould, Steven A. 44
Grassley, Charles 42
Great Britain 156
Great Depression 2, 3
gross domestic product (GDP) 136, 138

H

Habtegiris, Tirhas 101, 104–106, 179*c*
Harlow, Harry 54, 58–61, 174*c*
Harvard Ad Hoc Committee to Examine the Definition of Brain Death 82–84, 101
Harvard University 55–57, 175*c*
health
    defining 158–160, 168
    enhancement v. 168–172
"Health and Creative Adaptation" (Dubos) 168
Health and Human Services, U.S. Department of 11
health care 136–138, 137*t*, 138*t*. *See also* costs of health care
healthy volunteers 10
heart-lung death 79, 84–85, 116
heart transplants 115, 177*c*
Helms, Jesse 61, 176*c*
helper T-cell. *See* CD4 cell
hemorrhagic shock 38
hepatitis 25–27, 162, 163, 174*c*
HIV (human immunodeficiency virus) 147–153, *150*, *153–155*
    approaches to prevention 154, 156
    and cross-species infection 67
    U.S. funding for treatment/prevention 148–149
    vertical transmission trials 15–22, *16*, 175*c*
HIV-1 M 147
HIV-2 147
HIV/AIDS pandemic 15, 146*m*, 146–156, *150*, *153–155*
Ho, David 147
Holmesburg prison 24
Holocaust 8–10, *9*
homosexuality 160
hospice 132–134, *133*, 177*c*
Hounsfield, Godfrey Newbold 109, 177*c*
"How Much of the Brain Must Be Dead?" (Brody) 80
human enhancement 169–173
human immunodeficiency virus. *See* HIV
hyperactive children 170–171
hyperacute rejection 64, 66

I

IACUC. *See* Institutional Animal Care and Use Committee
Ibsen, Bjørn xiv, *xv*, 99–100, 176*c*
ICI Pharmaceuticals Division 102
Immerge BioTherapeutics 66, 68
immune response 64–66
immunosuppressive drugs 67, 115, 177*c*
Imutran 70, *71*
infant mortality 127–128
infection 68
informed consent
    in cross-cultural setting 18–19
    emergency room patients 38
    HIV vertical transmission trials 19
    plutonium injection experiments 36
    U.S. regulations 11
    Willowbrook hepatitis study 27
inner-city children 175*c*
insects, experimentation on 50
Insel, Thomas R. 171
Institute for OneWorld Health 145
Institutional Animal Care and Use Committee (IACUC) 61–62
Institutional Review Boards (IRBs) 11–13, 12*t*, 45, 175*c*

insurance
    for children 139, 140, 156–157
    Tirhas Habtegiris case 105
    for mental health conditions 166–168
    and rising health care costs 138–139
integrase inhibitors 152, *153*
intellectual property 55–57
ionizing radiation 35
IRBs. *See* Institutional Review Boards
iron lung 99, *100*
irreversible brain damage 108
irreversible coma 82–83, 100, 177c

J

jejunostomy tube 106
John Paul II (pope) 107
Jones, Leodus 24
Jordan, V. Craig 103–104
Judeo-Christian tradition 52
justice 12, 14

K

Kass, Leon 125
Kennedy, Edward M. 23, 24, *167*, 175c
Kennedy Krieger Institute (KKI) 29–34, *30*, 175c, 176c
Kenyon, Cynthia 124
kidney xenotransplants 63, 64, 174c
Killen, Jack 21
King, Nancy M. P. 43
Kipnis, Kenneth 42–43
KKI. *See* Kennedy Krieger Institute
Klintmalm, Goran B. 85
Krugman, Saul 25
Kübler-Ross, Elisabeth 132–133

L

Lauterbur, Paul 110, 178c
LD_{50} (lethal dose, 50 percent) test 49
lead-based paint 29–34, *30*, 33t, 175c, 176c
lead poisoning 29–32, *30*

learned helplessness *60*
Leavitt, Michael 42
Lee, Barbara Combs 91
life expectancy 122–123, 126–128, 134
life-extending technology 98–121
    feeding tube 106–109
    mechanical ventilator 98–101, *99, 100,* 104–106
    neuroimaging 109–114, *113*
    organ transplantation 114–120, 115t, 117t, *118*
life extension 122–135
    chronic disease and 129–131, 130t, 131t
    ethical positions on 125–126
    long-term care 128, 132
    science of *123,* 123–124
lifestyle diseases 129–131
liver damage 25
liver transplants for alcoholics 161–163
living donors 116
living wills 90–94
Lock, Margaret 81
locked-in syndrome 88, 90, 176c
London, Alex John 34
"Longevity, Identity, and Moral Character" (Overall) 125
long-term care 128, 132
low-income neighborhoods 29–32, *30*
Lucas, James B. 6
Lurie, Peter 20
Lyons, Dan 72

M

magnetic resonance imaging (MRI) 110, 178c. *See also* fMRI
malaria 145
Malley, Paul 91
manic-depressive illness 164
Mansfield, Sir Peter 110, 178c
Mason, Bill 54, 60
mechanical ventilator *99, 100*
    Tirhas Habtegiris case 104–106
    invention of 98–101, 176c

and organ transplantation 115
Karen Quinlan case 75, 76
Medicaid 105, 116, 139
medical imaging 109–111. *See also
specific types of imaging, e.g.* fMRI
medical proxies 91–94
medical research, animals for. *See
animal research*
Medicare 107, 116
Memorandum of Understanding
(MOU) 62
mental health parity 163–168, 165*t*,
167
mental retardation 174*c*
metastatic cancer 101
mice 50
microbicide 149
minimally conscious 88–90, 178*c*
minimal risk 28
monkeys 147
Moodley, Keymanthri 19
Moss, Alvin H. 163
mother-to-child transmission. *See
HIV, vertical transmission trials*
MOU (Memorandum of
Understanding) 62
MRI (magnetic resonance imaging)
110, 178*c*. *See also* fMRI
Munson, Ronald 28, 79–80
myelogenous leukemia 118

N

Nakajima, Hiroshi 130–131
nasogastric (NG) tube 106
National Commission for the
Protection of Human Subjects
of Biomedical and Behavioral
Research 6, 11, 175*c*
National Institute on Aging (NIA)
123
Nazism 8–10, *9*
needle-exchange programs 156
Nelson, Robert M. 43, 45
neuroimaging 109–114, *113*,
120–121. *See also specific types of
neuroimaging, e.g.:* fMRI
Newman, Russ 166

Nexium 142
NG (nasogastric) tube 106
NIA (National Institute on Aging)
123
Nipah virus 67
nonhuman primates (NHP)
47–48
nontherapeutic experiment 2
Nordal, Katherine 164, 168
Northfield Laboratories 39–41, 43,
44, 176*c*
Nuremberg Code 8–10, *9*, 174*c*
nursing-home care 132

O

Oak Ridge, Tennessee 36, 174*c*
Office for Human Research
Protections (OHRP) 13, 42, 45
OncoMouse 55–57, *56*, 175*c*
oppositional defiant disorder 170
ordinary care 87, 107
organ donor 81–85
organism (term) 79
organ transplantation
conceiving a child for *118*,
118–120
as life-extending technology
114–120, 115*t*, 117*t*, *118*
xenotransplantation 63–72, *65*,
71
Overall, Christine 125
Owen, Adrian M. 111
oxygen
and mechanical ventilator 99,
100
PolyHeme trial 39
Theresa Schiavo case 88

P

pain 86
palliative care 104–105, 134
Palm Sunday Compromise 95
Pappworth, M. H. 27
Parens, Eric 172–173
parental consent 28, 31–32
patent issues 55–57, 142, 175*c*
paternalism 10

Paul Wellstone Mental Health
  Equitable Treatment Act 168
PEG (percutaneous endoscopic
  gastrostomy) tube 106–107,
  177c
Pence, Gregory 32
penicillin 2, 5
PEPFAR (President's Emergency
  Plain for AIDS Relief) 148
persistent vegetative state (PVS)
  and brain death definitions 83
  Cruzan v. Missouri 108
  Theresa Schiavo case 88, 89
personhood 83
PERV (porcine endogenous
  retroviruses) 68–69
physician-assisted suicide 85–86
pigs 63, 67–69
Pius XII (pope) 176c
placebo 17, 18, 21
plutonium 35–37, 174c
polio xiv
Pollard, Charles 4
PolyHeme synthetic blood trial
  38–45, *40*, 176c
porcine endogenous retroviruses
  (PERV) 68–69
Post, Stephen 107
poverty 29–32, *30*
PPL Therapeutics 66
Pravachol 142, 144–145
premature burial 76, 77
President's Emergency Plain for
  AIDS Relief (PEPFAR) 148
Prilosec 142
prisoners, experiments on 24
prolongevity 123
protease inhibitors 152, *153*
protocol 6, 39
public health 66–70
Public Health Service, U.S. 2, 3,
  61–62
PVS. *See* persistent vegetative state

Q

Quinlan, Karen Ann 74–76, 83, 87,
  95–96, 177c

R

rabbits 50
Rachels, James 119–120
radiation experiments 35–37, 174c
radiation therapy 101
*Raising the Dead: A Doctor's Encounter
  with His Own Mortality* (Selzer)
  80–81
*Raising the Dead: Organ Transplants,
  Ethics, and Society* (Munson) 79–80
rats 50
Reemtsma, Keith 63, 174c
Regan, Tom 52–53
replacement, reduction, and
  refinement 62
research, animal. *See* animal
  research
research and development 142
respect for persons
  Belmont Report 11, 14
  HIV vertical transmission trials
    18–19
  IRBs 12
retrovirus 68, 151
reverse transcriptase inhibitor 152,
  153
rhesus monkey 54, 58–59, 61
ribonucleic acid (RNA) 151
right of privacy 75–76, 177c
right-to-die/right-to-live cases
  Cruzan v. Missouri 108–109
  Tirhas Habtegiris case 101,
    104–106
  Karen Quinlan case 74–76,
    95–96
  Theresa Schiavo case 87–90,
    94–96
RNA (ribonucleic acid) 151
robotics 176c
Roman Catholicism 87, 107–108
Rowan, Andrew 50–54, 61–63
Russell, William 62

S

Saba, Joseph 17
St. Christopher's Hospice (London)
  132, *133*, 177c

Salomon, Daniel 68
salvarsan (arsphenamine) 5, 102
Salvi, Daniel 105
Saunders, Cicely 132, *133*, 177*c*
Saviola, Marilyn 91
Scalia, Antonin 77–78
Schatz, Irwin J. 3
Schiavo, Michael 87–88, 94–95
Schiavo, Theresa 83, 87–96, 105, 178*c*
Schindler, Robert and Mary 94–95
SCHIP. *See* State Children's Health Insurance Program
Schneider, Carl E. 92, 93
Scott, Lester 4
Selzer, Richard 80–81
sentience 53
Shalala, Donna 20
Shark Fin Project 142
Shaw, Herman 7
Shorett, Peter 57
shyness 170–171
Siegler, Mark 163
Singer, Peter 51–52, 59–60, 62, 175*c*
"Social and Justice Implications of Extending the Human Life Span" (Chapman) 125
Social Darwinism 159–160
sooty mangabey 147
speciesism 51, 52
spinal tap 3
spirochete 5
standard of care 15, 19–21
Starzl, Thomas 69
State Children's Health Insurance Program (SCHIP) 139, 140, 156–157
stem cells 117
stigmas 154, 161, 165–166
Stock, Gregory 134–135
stroke 142, 144, 145
suffering 50–51, 70, 71, 72–73
suicide 165
Supreme Court, Canada 57
Supreme Court, Florida 95
Supreme Court, Missouri 108–109
Supreme Court, New Jersey 177*c*
Supreme Court, U.S.
  Cruzan v. Missouri 109
  OncoMouse 55
  Theresa Schiavo case 95
  terminal sedation decision 86
Sus scrofa domesticus 67
synthetic blood substitute. *See* PolyHeme synthetic blood trial

T

tamoxifen 102–104, *103*
technology, health care costs and 140–146, 141*t*, 143*t*, 144*t*
terminal illness 35
terminal sedation (TS) 86–87
"Terri's Law" 95, 178*c*
Thailand 156
Tong, Rosemarie 169
toxicity tests 49
tracheostomy tube 99
transgenic animals 55–57, *56*, 66, 175*c*
transhumanism 169–172
transplantation
  and death of donor 81–84
  first heart transplant 177*c*
  liver transplants for alcoholics 161–163
  xenotransplantation 63–72, *65*, *71*
triple cocktail 146, 152, 178*c*
TS (terminal sedation) 86–87
tube feeding. *See* feeding tube
tumor 104
Tuskegee syphilis study 1–8, *4*, 7, 21, 160, 174*c*, 175*c*
Twice Dead: Organ Transplants and the Reinvention of Death (Lock) 81

U

Uganda 154, 156
Ullmann, Emerich 64
Uncaged Campaigns 70, *71*, 176*c*
Uniform Determination of Death Act (UDDA) 84
uninsured 139

United States
  animal research in 46–47
  health care costs 136–139
  life expectancy 126–127
universal health care 138
Upjohn (pharmaceutical company) 56
utilitarianism 51–52, 56

V
vaccines 151
Vaslef, Steven 39
vegetative state 111–114, *113*, 177*c*, 179*c*
ventricular tachycardia 80–81
vertical chamber apparatus 59
viral encephalitis 67
viruses 67, 69
visceral leishmaniasis (VL) 145
voluntary consent 9, 174*c*
vulnerable populations 23–45
  animals as 46
  and IRBs 13
  Kennedy Krieger Institute lead paint study 29–32, *30*
  plutonium injection experiments 34–37
  PolyHeme synthetic blood trial 38–45, *40*
  protections for emergency room patients 37–38
  Willowbrook hepatitis study 25–27

W
Wada, Juro 177*c*
waiting lists, organ transplant 115*t*, 117*t*

waived-consent trial 38
Wallis, Terry 89, 178*c*
Wall Street Journal 40, 41
Walpole, Arthur L. 102, 103
war crimes 8–10, *9*
Weiss, Robin 68
wellness 172
WHO (World Health Organization) 168, 169
whole-brain death 79, 83, 84
whole-organ transplants 63
Williams, J. D. 4–5
Willowbrook hepatitis study 25–27, 174*c*, 175*c*
Wilson, Cecil B. 91
Wolfson, Jay 94, 95
women, 19th century medical attitudes towards 158–159
Wong, Mitchell 128
World Health Organization (WHO) 168, 169
World Medical Association 10, 174*c*
World War II 8–10, *9*, 34

X
xenotransplantation 63–73
  as alternative to human donors 117
  animal suffering 70, *71*, 72
  first modern experiments 63, 174*c*
  public health risks 66–70
  rejection 64–66, *65*

Z
ZDV. *See* AZT